PROTEIN BIOCHEMISTRY, SYNTHESIS, STRUCTURE AND CELLULAR FUNCTIONS

α-SYNUCLEIN

FUNCTIONAL MECHANISMS, STRUCTURE AND ROLE IN PARKINSON'S DISEASE

PROTEIN BIOCHEMISTRY, SYNTHESIS, STRUCTURE AND CELLULAR FUNCTIONS

Additional books in this series can be found on Nova's website under the Series tab.

Additional e-books in this series can be found on Nova's website under the e-book tab.

PROTEIN BIOCHEMISTRY, SYNTHESIS, STRUCTURE AND CELLULAR FUNCTIONS

α-SYNUCLEIN

FUNCTIONAL MECHANISMS, STRUCTURE AND ROLE IN PARKINSON'S DISEASE

MARK POLIZZI
AND
HOWARD C. KANOWITZ
EDITORS

NOVA BIOMEDICAL

New York

For permission to use material from this book please contact us:
Telephone 631-231-7269; Fax 631-231-8175
Web Site: http://www.novapublishers.com

Library of Congress Cataloging-in-Publication Data

Library of Congress Control Number: 2013952268

ISBN: 978-1-62948-493-8

Published by Nova Science Publishers, Inc. † New York

Contents

Preface

α -synuclein is a synuclein protein of unknown function primarily found in neural tissue, making up as much as 1% of all proteins in the cytosol. It is predominantly expressed in the neocortex, hippocampus, substantia nigra, thalamus, and cerebellum. In this book, the authors discuss the functional mechanisms, structure and role in Parkinson's disease of alpha-synuclein. Topics include α-Synuclein accumulation in Parkinson's disease and multiple system atrophy; α-Synuclein metabolism and aggregation in the pathogenesis of Parkinson's disease; and the potential role of α-Synuclein dysregulation in Parkinson's disease pathology.

Chapter I – α-Synuclein is one of the proteins fundamental to the pathogenesis of neurodegenerative diseases. This protein forms filamentous inclusions in various neurodegenerative diseases, which are known as synucleinopathies. Synucleinopathies are pathologically classified into two groups: Lewy body diseases and multiple system atrophy (MSA). Lewy body diseases, including Parkinson's disease (PD) and dementia with Lewy bodies, exhibit the common pathological feature of neuronal cytoplasmic inclusions called Lewy bodies. α-Synuclein has been identified as the main component of Lewy bodies, and a missense mutation in the α-Synuclein gene is the genetic cause of familial parkinsonism. Moreover, evidence implicates α-Synuclein and its accumulation in PD pathogenesis. In contrast, MSA is an intractable neurodegenerative disease caused by α-Synuclein accumulation in oligodendrocytes and neurons. The pathological hallmark of MSA is the presence of glial cytoplasmic inclusions, which are mostly found in oligodendrocytes and are composed of α-Synuclein. α-Synuclein accumulation plays an important role in neurodegeneration observed in both PD and MSA and may be the primary lesion that eventually compromises nerve cell function and viability. However, the cellular mechanisms underlying each of these neurodegenerative diseases seem to be different. This chapter introduces recent progress in the understanding of neurodegeneration mechanisms in PD and MSA and focuses on studies of neuropathology, molecular genetics, animal models, and cellular mechanisms.

Chapter II – α-Synuclein (α-Syn) is a major constituent of Lewy bodies, the pathological intracellular inclusion bodies of Parkinson's disease (PD), dementia with Lewy bodies, and multiple system atrophy. The intracellular aggregates or the accumulation of toxic oligomer of α-Syn is implicated in the pathogenesis of these diseases. Various forms of α-Syn, such as monomer, oligomer, and fibril, are metabolized by the ubiquitin-proteasome system, autophagy-lysosome pathway, chaperone-mediated autophagy, and other proteolytic systems. The impairment of these systems leads to the accumulation of α-Syn, resulting in neurodegeneration. In this article, the authors focus on the current knowledge about the involvement of α-Syn metabolism in PD pathogenesis and discuss proteolytic systems as a therapeutic target for PD.

Chapter III – The motor symptoms of Parkinson's disease (PD) arise after a marked loss of substantia nigra dopaminergic neurons. Common to the surviving nigral neurons in PD brain are proteinaceous aggregates known as Lewy bodies, which contain high levels of the presynaptic protein α-Synuclein. Despite being strongly implicated in PD, the precise role of α-Synuclein in PD neuropathology remains inconclusive. However, it is known that the α-Synuclein protein contributes to the normal functioning of dopaminergic neurons in part by interacting with a number of key regulatory proteins and other cellular components. Normally, α-Synuclein is present at low levels and in a soluble form that can interact with proteins to modulate their functions, especially those that regulate dopamine production, release, and reuptake. In aging human brain, α-Synuclein levels rise, which triggers over-modulation of α-Synuclein interacting proteins. Eventually, high levels of α-Synuclein lead to: (i) Lewy body pathology, (ii) α-Synuclein accumulation in insoluble oligomers that impairs its normal regulatory functions, (iii) dysregulation of dopaminergic cellular function, (iv) the death of nigral dopaminergic neurons, and (v) the emergence of PD motor symptoms. This commentary will describe normal functions of α-Synuclein in dopaminergic neurons, and explore how changes in the levels and solubility of intraneuronal α-Synuclein may contribute to PD.

In: α-Synuclein ISBN: 978-1-62948-493-8
Editors: M. Polizzi and H. C. Kanowitz © 2014 Nova Science Publishers, Inc.

Chapter I

α-Synuclein Accumulation in Parkinson's Disease and Multiple System Atrophy

Ikuru Yazawa* and Yasuyo Suzuki
Laboratory of Research Resources, Research Institute,
National Center for Geriatrics and Gerontology, Aichi, Japan

Abstract

α-Synuclein is one of the proteins fundamental to the pathogenesis of neurodegenerative diseases. This protein forms filamentous inclusions in various neurodegenerative diseases, which are known as synucleinopathies. Synucleinopathies are pathologically classified into two groups: Lewy body diseases and multiple system atrophy (MSA). Lewy body diseases, including Parkinson's disease (PD) and dementia with Lewy bodies, exhibit the common pathological feature of neuronal cytoplasmic inclusions called Lewy bodies. α-Synuclein has been identified as the main component of Lewy bodies, and a missense mutation in the α-Synuclein gene is the genetic cause of familial parkinsonism. Moreover, evidence implicates α-Synuclein and its

* Corresponding author: Ikuru Yazawa, MD, Ph.D., Laboratory of Research Resources, Research Institute, National Center for Geriatrics and Gerontology, 35 Gengo, Morioka-cho, Obu-shi, Aichi 474-8511, Japan, Phone: +81-562-46-2311, Fax: +81-562-46-8319, E-mail: yazawaik@ncgg.go.jp.

accumulation in PD pathogenesis. In contrast, MSA is an intractable neurodegenerative disease caused by α-Synuclein accumulation in oligodendrocytes and neurons. The pathological hallmark of MSA is the presence of glial cytoplasmic inclusions, which are mostly found in oligodendrocytes and are composed of α-Synuclein. α-Synuclein accumulation plays an important role in neurodegeneration observed in both PD and MSA and may be the primary lesion that eventually compromises nerve cell function and viability. However, the cellular mechanisms underlying each of these neurodegenerative diseases seem to be different. This chapter introduces recent progress in the understanding of neurodegeneration mechanisms in PD and MSA and focuses on studies of neuropathology, molecular genetics, animal models, and cellular mechanisms.

Introduction

Recent advances in molecular biology and genetics have provided important insights into the etiology and pathogenesis of neurodegenerative diseases characterized by the progressive dysfunction and death of neurons. The understanding of the fundamental genetic and molecular basis of neurodegenerative diseases has led to the development of tools used to generate many disease models. Traditional morphological and immunohistochemical approaches in neuropathology are not only essential to the establishment of a diagnosis at autopsy but also contribute to the understanding of the pathogenic mechanisms of neurodegeneration. Static evaluation based on morphological observation has provided new insights into the dynamic pathophysiology of neurodegeneration.

α-Synuclein has been identified as one of the proteins fundamental to the pathogenesis of neurodegenerative diseases [1]. In addition, α-Synuclein accumulates intracellularly as inclusion bodies in the central nervous system (CNS). Cellular inclusions have long been recognized as being characteristic of certain degenerative processes and are required features for the histological diagnosis of a specific disease [2]. Although α-Synuclein accumulation is a common primary degenerative process in a certain group of neurodegenerative diseases, unique morphological features suggest that the mechanisms underlying α-Synuclein accumulation are not identical among these diseases. Recent genetics research has shown that Parkinson's disease (PD) is actually not a single disease entity; rather, it seems to be a heterogeneous group of diseases associated with an overlapping spectrum of clinical and pathological changes [3]. Correspondingly, histological studies revealed the different features of α-Synuclein inclusions among neurodegenerative diseases and suggested that independent pathological mechanisms underlying α-Synuclein accumulation cause respective neurodegeneration. This chapter provides an overview of the current understanding of the correlation between histological and molecular studies of neurodegenerative diseases, with a particular focus on the characteristic α-Synuclein accumulation as inclusion bodies in PD and multiple system atrophy (MSA). The potential role of α-Synuclein is critically discussed, with an emphasis on how α-Synuclein accumulation causes neurodegeneration. A better understanding of the processes of these diseases will be promising for future therapeutic approaches.

Synucleins and Synucleinopathies

Synucleins are proteins that are abundant in CNS. The synuclein family consists of three distinct genes for α-Synuclein, β-synuclein, and γ-synuclein. The first sequence of α-Synuclein was obtained from the electric organ of the Pacific ray (*Torpedo californica*) [4]. Later, α-Synuclein was identified as the non-amyloid β component precursor in amyloid plaques in the brains of patients with Alzheimer's disease (AD) [5]. A series of studies demonstrated that α-Synuclein is localized to presynaptic terminals and nuclei in CNS neurons [4, 6-9]. α-Synuclein contains 140 amino acids. Human α-Synuclein and β-synuclein are more closely related to each other than to γ-synuclein [10]. Synucleins have only been identified in vertebrates [11]. Although the function of α-Synuclein is not understood, synucleins have a modulatory, rather than an essential, synaptic function [10]. α-Synuclein knockout mice demonstrated that α-Synuclein is not essential for neuronal development or for the formation of presynaptic terminals; however, α-Synuclein modulates dopamine release in the striatal brain [12]. Synucleins are natively unfolded proteins, i.e., they have no secondary or tertiary structure in solution, and hence are heat stable [6, 10, 13]. However, after binding to lipid membranes or detergent micelles, α-Synuclein undergoes a structural transition to an α-helical conformation [14-17]. Human α-Synuclein readily oligomerizes to higher-order species, initially to dimers, subsequently to trimers, and ultimately to oligomers, culminating in large fibrils or amyloids [10]. *In vitro* studies demonstrated that α-Synuclein mutations in PD influence fibril formation [18-21]. Filaments composed of recombinant monomeric human α-Synuclein have the same morphological and ultrastructural characteristics as disease filaments [11]. Mutation in the human α-Synuclein (*SNCA*) gene increases the rate of α-Synuclein assembly [18, 20, 22]. Post-translational modifications can also accelerate fibril formation. α-Synuclein is phosphorylated, most prominently at Ser 129, which increases its ability to aggregate and form fibrils [23]. α-Synuclein accumulation is associated with various neurodegenerative diseases, which are known as synucleinopathies. These include PD, dementia with Lewy bodies (DLB), AD, Down syndrome, neurodegeneration with brain iron accumulation (previously termed Hallervorden–Spatz disease), and MSA. The clinical phenotype of synucleinopathies includes movement disorder, dementia, and various peripheral nerve system disorders [24]. α-Synuclein plays a crucial role in neurodegeneration development in PD and other synucleinopathies.

Lewy Bodies and Other Types of α-Synuclein Accumulation

The pathological hallmark of PD is the presence of characteristic neuronal inclusions, termed Lewy bodies, accompanied by the loss of dopaminergic neurons in the substantia nigra. Lewy bodies are concentric hyaline neuronal cytoplasmic inclusions originally described in PD [24]. Frederich Lewy first described the characteristic inclusions associated with PD [25]. Tretiakoff named the inclusions "corps de Lewy" (Lewy bodies) and recognized their presence in the substantia nigra in autopsy cases of PD [26]. Histologically, there are two types of Lewy bodies: classical and cortical types [2]. Classical Lewy bodies are neuronal cytoplasmic inclusions that are usually spherical and have an eosinophilic core surrounded by a paler halo in hematoxylin–eosin staining (Figure 1A). They are found in the substantia nigra and other pigmented brain stem nuclei. Cortical Lewy bodies are found in neurons located in the deeper part of the cerebral cortex [2]. These are less clearly defined than classical Lewy bodies and consist of a homologous zone of hypereosinophilia that usually lacks a surrounding halo (Figure 1B). Cortical Lewy bodies are present in small numbers in all PD cases but are usually more extensively distributed in DLB cases [2]. DLB is clinically characterized by fluctuation, early visual hallucinations, and tremor. DLB patients typically have Lewy bodies in the brain stem and midbrain nuclei, as well as a widespread distribution of these bodies in the neocortex and limbic cortex [2]. The presence of Lewy bodies defines several conditions, which are termed Lewy body diseases, e.g., PD, PD with dementia, and DLB. In aged brains, Lewy bodies may be found in the absence of a defined clinical syndrome or without any overt neurological disease [24]. It has been suggested that such individuals have presymptomatic PD; however, incidental Lewy bodies do not imply that all such individuals are on an inevitable trajectory toward PD [24, 27].

α-Synuclein is a major component of Lewy bodies in PD and DLB [28]. Lewy bodies are immunoreactive for ubiquitin, indicating that α-Synuclein itself is ubiquitinated or that Lewy body formation precedes ubiquitination [1]. Lewy bodies also contain many other proteins. However, it is generally believed that aggregation of highly ordered α-Synuclein fibrils is crucial to the disease process [29], although its precise relationship with neuronal damage remains controversial. There is much debate in the literature over whether Lewy bodies themselves are toxic to dopaminergic neurons [10]. The relevance of protein accumulation for a neurodegenerative process in other types of neurodegenerative diseases remains under debate. For instance, the relevance of neuronal intranuclear and cytoplasmic inclusions, which are also immunoreactive to ubiquitin, in the neurodegenerative process of polyglutamine diseases is under discussion [1]. The polyglutamine diseases are a group of hereditary neurodegenerative disorders caused by the expansion of CAG trinucleotide repeats that encode a polyglutamine tract in the genes involved [30]. Aggregation of gene products that carry an expanded glutamine repeat is believed to be a primary pathological mechanism in polyglutamine diseases. However, the onset of a neurological phenotype or cellular dysfunction mediated by the expanded polyglutamine repeats in the responsible gene product was independent of the formation of inclusions [31-33]. Thus, the relationship between neuronal inclusions and neurotoxicity in polyglutamine diseases remains controversial [34]. Soluble polyglutamine oligomers have been proposed as the cytotoxic structure of proteins with expanded polyglutamine repeats [35, 36]. Similarly, in Lewy body diseases, it is hypothesized that Lewy body formation is a neuroprotective reaction [37] and that soluble α-Synuclein oligomers cause neurodegeneration. Conversely, Lewy bodies may themselves have the deleterious effect of space-occupying nuisance in neurons. Lewy bodies represent α-Synuclein accumulation that is too large for neurons, leading to neuronal cell death.

An alternative cause of parkinsonism needs to be pursued in autopsy cases that exhibit cell loss from the pigmented neuronal nuclei in the midbrain and brain stem in the absence of Lewy bodies. In drug-induced parkinsonism, post-encephalitic parkinsonism, and juvenile forms of autosomal recessive parkinsonism, although neuropathological examination has revealed nigrostriatal degeneration in a pattern closely resembling that of idiopathic PD, there is cell loss without evidence of Lewy body or α-Synuclein pathology [38]. It is not known if Lewy bodies contribute to neurodegeneration or are a protective mechanism to slow neuronal death. However, it is clear that the formation of Lewy bodies is a disease process of α-Synuclein accumulation, and that the absence of Lewy bodies implies a process that is different from that occurring in the presence of Lewy bodies in neurodegeneration, which is involved even in the same clinical group of parkinsonism.

Another significant morphological feature of α-Synuclein accumulation is the occurrence of distinctive neuritic degeneration in synucleinopathies. Immunocytochemical staining of α-Synuclein has revealed abnormal neurites in areas undergoing neurodegeneration with Lewy bodies [24]. In MSA, anti-α-Synuclein antibody detected widespread neuropil threads in the affected lesions that were accompanied by the characteristic oligodendrocytic inclusions; however, the antibody did not stain Lewy bodies. Although both PD and MSA are principally caused by α-Synuclein accumulation, a significant difference exists in the pathological features of the two diseases. Therefore, it is speculated that neuronal accumulation of α-Synuclein occurs in different pathways that cause neurodegeneration in the two diseases.

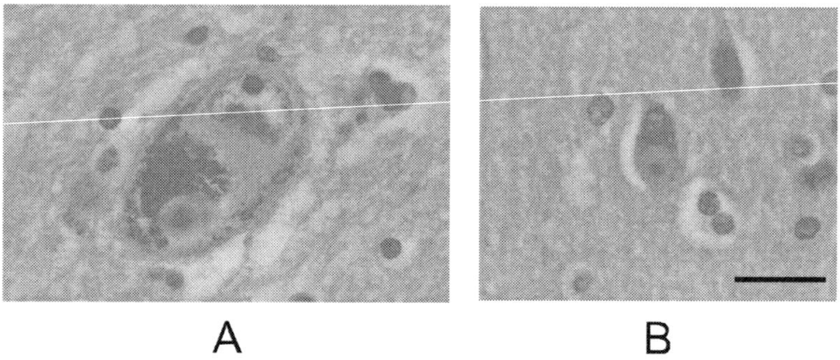

Figure 1. A. Classical Lewy body in a pigmented neuron of the substantia nigra in a patient with PD. The Lewy body has a hyaline eosinophilic core with a pale halo. B. Cortical Lewy body in a patient with DLB does not have a distinct halo. Hematoxylin and eosin staining. Scale bar = 20 μm.

α-Synuclein and PD

Primary parkinsonism is known as PD and can be sporadic or familial. It is the most common type of parkinsonism and the second most common neurodegenerative disease [38]. The core biochemical pathology in parkinsonism is decreased dopaminergic neurotransmission in basal ganglia [38]. PD is clinically characterized by three features: bradykinesia, rigidity, and resting tremor. The diagnosis is based on the presence of two of these features. On average, familial parkinsonism has a younger age of disease onset than sporadic parkinsonism. Although the diagnosis is based on these clinical features of parkinsonism, the criteria necessary for the definitive diagnosis of PD include post-mortem confirmation [39]. The diagnostic hallmarks of idiopathic PD are the presence of Lewy bodies and cell loss in the substantia nigra and other pigmented brain stem nuclei. Cell loss is accompanied by astrocytes and macrophages. The neuroanatomical regions that exhibit α-Synuclein accumulation in idiopathic PD follow a consistent pattern as the disease progresses [40]. The pathology is initially confined to the olfactory bulbs and caudal brain stem, especially the dorsal motor nucleus of the vagus in the medulla. The pathology then spreads upward to include neuritic changes in the brain stem and to the telencephalon and cerebral cortex. The final involvement of the cerebral cortex is consistent with the potential to develop dementia late in the disease course [38].

Two discoveries verified that α-Synuclein contributes to PD pathogenesis: the identification of α-Synuclein as a major component of Lewy bodies and the discovery that a missense mutation in the α-Synuclein (*SNCA*) gene is the genetic cause of familial parkinsonism [1]. Three independent missense point mutations in *SNCA*, namely A53T, A30P, and E46K, have been linked to the early-onset familial forms of parkinsonism [41-43]. Despite the variability in clinical pictures, cases with *SNCA* mutations pathologically show severe cell loss in the pigmented brain stem nuclei in association with Lewy bodies [24]. The amino acid located at position 53 in the mouse and rat α-Synuclein is a threonine, not an alanine. This suggests that it is not simply the presence of a threonine at position 53, but some difference in the conformation of the mutant protein, which is specific to human, but not rodent, α-Synuclein, that is pathogenic [11]. An even more direct causative link between α-Synuclein and PD at the genetic level was provided by the more recent discovery that not only point mutations but also duplications and triplications of the entire wild-type sequence of *SCNA* cause autosomal dominant parkinsonism with extensive α-Synuclein inclusions [44, 45]. The clinical and pathological phenotypes associated with the duplication and triplication of *SNCA* resemble Lewy body diseases. Overexpression of wild-type α-Synuclein has been identified as a cause of PD and DLB [11], and the identification of the *SNCA* mutations suggests that a mere increase in α-Synuclein protein levels can be directly toxic to neurons [3].

In addition to the *SNCA* mutations, many monogenic forms of familial parkinsonism associated with various mutations have been identified; however, their pathogenic mechanisms remain unclear. For instance, mutations in the gene that encodes the leucine-rich repeat kinase 2 (*LRRK2*) are associated with a substantial proportion of cases of dominantly inherited parkinsonism [46]. Mutations in the *parkin* and PTEN-induced kinase 1 (*PINK1*) genes [47, 48] cause autosomal recessive parkinsonism, and *DJ-1* mutations cause other rare forms of familial parkinsonism [49]. Parkin is an E2-dependent E3 ubiquitin–protein ligase, and mutations in the *parkin* gene result in juvenile parkinsonism that is slowly progressive, with an onset usually before age 40 [38, 47]. Neurodegeneration in the substantia nigra is observed in juvenile parkinsonism; however, no Lewy bodies are found in most cases [50]. Impairment of the ubiquitin–proteasome system can result in neuronal accumulation of proteins. However, double transgenic mice overexpressing mutant parkin and mutant α-Synuclein showed that the function of parkin does not have a significant impact on α-Synuclein processing [51]. Parkinsonism associated with *parkin* mutations involves a mechanism that is independent of that underlying parkinsonism associated with *SNCA* mutations [52], which is consistent with the neuropathological findings of neurodegeneration in the absence of Lewy bodies. *PINK1* encodes a putative serine/threonine kinase with a mitochondria-targeting sequence, indicating localization to the mitochondria [48]. Oxidative stress can be linked to mitochondrial dysfunction and has long been considered a prominent pathogenic mechanism in PD because of some properties of substantia nigra neurons, including the observation that the neuromelanine pigment and high iron content lead to the formation of reactive oxygen species [53]. The expression of the *PINK1* mutation in a *Drosophila* model resulted in defects in mitochondrial morphology, reflected through fragmented mitochondrial cristae, and increased sensitivity to multiple stresses, including oxidative stress [54, 55]. The pathogenic mechanism of parkinsonism most probably involves mitochondrial dysfunction through oxidative stress. Together, evidence suggests that idiopathic PD is not the result of a single causative factor; rather, it seems to be multifactorial, integrating many cellular systems to work in concert to induce progressive degeneration in vulnerable neuronal populations [53].

Animal Models of PD and Cellular Mechanisms of Neurodegeneration

The identification of *SNCA* mutations that cause familial parkinsonism has led to the generation of many animal models of PD (overexpressing wild-type and mutant α-Synuclein). Experimental animal models of PD are essential for studying disease pathogenesis and for identifying approaches aimed at interfering with the disease process. These models have been generated in various organisms. Although the invertebrate genome does not encode α-Synuclein, *Drosophila* models have proven to be extremely useful in the study of α-Synuclein neurotoxicity [3]. Overexpression of human wild-type and mutant α-Synuclein in *Drosophila* resulted in an age-dependent specific degeneration of dopaminergic neurons, a motor phenotype, and α-Synuclein-positive intracytoplasmic inclusions [56]. The models have successfully replicated PD characteristics, namely neuron selectivity, α-Synuclein pathology, and age dependence. These findings are compatible with the hypothesis that the ordered assembly of α-Synuclein into filaments is both necessary and sufficient for the degeneration of dopaminergic nerve cells [1]. Because an increased number of inclusion bodies are correlated with reduced toxicity, these experiments have been interpreted to support the hypothesis that inclusion bodies protect neurons from α-Synuclein toxicity [37]. The murine models, depending on the promoter used, show late-onset neuronal cell death and similar α-Synuclein inclusions. One study demonstrated the presence of α-Synuclein filaments in the brain and spinal cord of mice transgenic for human A53T α-Synuclein [57]. The formation of inclusions correlated with the appearance of a movement disorder. In addition to transgenic animals, viral expression systems have been used to introduce *SNCA* into specific regions of the midbrain [10]. The lentiviral-based expression of human α-Synuclein in the rat substantia nigra resulted in selective dopaminergic neuronal toxicity, with the formation of α-Synuclein oligomers rather than filaments [58].

The consistent pattern of accumulating α-Synuclein pathology in idiopathic PD raised the possibility that the disease process is initiated in a single cell, from where it may spread in a prion-like manner between synaptically connected neurons [11, 40]. The transmission of exogenous protein aggregates has been well studied in prion disorders [59]. In AD, which is associated with misfolded proteins such as β-amyloid and tau, it has been proposed that aggregates of misfolded proteins might propagate the misfolded state in a prion-like manner [60, 61]. The injection of brain extracts from neurofibrillary tangle-bearing mice into wild-type and mutant tau-expressing mice induced the seeding of tau fibrils in neurons [62, 63]. Synucleinopathies are no exception, and propagation was suspected in these diseases as well. Studies of the brains of PD patients who had received long-term fetal mesencephalic transplants showed the presence of Lewy bodies in grafts [64, 65]. These findings are consistent with the spread of seeds from the diseased host tissues to the grafts, followed by the nucleated assembly of α-Synuclein [11]. However, it remains unclear whether released α-Synuclein or a molecular signal is responsible for the development of α-Synuclein deposits in the transplanted cells. Experimental evidence of intercellular protein transfer and the seeding of aggregation has been obtained [11], and recombinant α-Synuclein was shown to be transmitted via endocytosis to neurons in culture [66]. A similar study showed that exogenous α-Synuclein fibrils induced the formation of Lewy body-like intracellular inclusions in α-Synuclein-overexpressing cells [67]. A single intrastriatal injection of synthetic α-Synuclein fibrils into wild-type mice led to neuron-to-neuron transmission of pathological α-Synuclein and consequent corruption of the endogenous protein [68]. There is growing evidence that cell–cell interactions contribute to the spread of neurodegeneration; however, the nature of these interactions remains unclear.

α-Synuclein and MSA

MSA is a neurodegenerative disease that affects oligodendrocytes and neurons in CNS [69]. MSA presents clinically with parkinsonism, ataxia, and autonomic failure [70]. MSA was previously believed to be three separate diseases known as striatonigral degeneration, olivopontocerebellar atrophy, and Shy–Drager syndrome. However, the recognition that glial cytoplasmic inclusions (GCIs) are the defining lesions in these three diseases led to the simplification of the classification of MSA patients to MSA-C (cerebellar ataxia) or MSA-P (parkinsonism), depending on the relative predominance of clinical and pathological abnormalities [71, 72]. MSA usually occurs in the sixth decade of life and is followed by relentless progression and death, which occurs after an average disease duration of 9 years [73]. No *SNCA* mutations have been detected in this disease [74]. The cellular mechanisms underlying neurodegeneration in MSA are not fully understood, and, unfortunately, no therapeutic approach has yet had an immediate impact on this disease.

MSA represents a group of disorders that exhibit diverse clinical and pathological manifestations [69]. The term is used to encompass a wide variety of cases with clinical and pathological features of striatonigral degeneration, olivopontocerebellar involvement, or autonomic failure [75]. CNS regions that are affected by the degenerative process include the cerebral and cerebellar white matter, cerebral and cerebellar gray matter, substantia nigra and locus ceruleus, putamen, pons, medulla oblongata, and spinal cord. Neuronal loss is most severe in the dorsolateral putamen, substantia nigra, locus ceruleus, Purkinje cell layer, basis pontis, inferior olivary nucleus, and nucleus of vagus and intermediolateral spinal column [76]. Neuronal loss is accompanied by variable astrocytosis in the affected regions. The location and severity of neuronal loss and astrocytosis in MSA are related to the clinical subtype and duration of illness [77]. For instance, the degree of neuronal loss in the substantia nigra and locus ceruleus is variable and generally correlates with the extent of parkinsonian features [75]. Interestingly, the remaining neurons do not show formation of Lewy bodies, which are the defining pathological hallmark of PD [24]. Although both PD and MSA are principally caused by α-Synuclein accumulation, a significant difference exists in the pathological features of the two diseases. Although loss of myelin staining in white matter tracts relates to the severity of neuronal loss in affected areas, widespread myelin damage can be identified in apparently normal areas in MSA [78]. This pathological finding indicates that degeneration initially develops in oligodendrocytes, leading to myelin disruption or neuronal degeneration. Some authors suggested that MSA shares traits with the family of demyelinating disorders that exhibit myelin disruption followed by axonal dysfunction [79, 80].

Three significant neuropathological features characterize MSA histologically: GCIs, neuronal inclusions, and neuropil threads; all three are composed of α Synuclein. GCIs, which were the first neuropathological manifestation described in MSA, are oligodendrocytic inclusions [81-83]. The presence of GCIs is noted as the consistent and unifying histopathological feature found in all of the various clinical and pathological manifestations of MSA [75]. In studies of GCIs, filaments isolated from the CNS of MSA patients were identified using anti-α-Synuclein antibodies [84]. α-Synuclein is a major component of inclusions in MSA [85, 86] and may be the primary lesion that eventually compromises nerve cell function and viability [1]. However, the relevance of oligodendrocytic α-Synuclein accumulation to neuronal degeneration in MSA remains unknown. The vulnerability of oligodendrocytes probably underlies the susceptibility to neuronal degeneration in MSA.

Animal Models of MSA

The understanding of the complexity of degeneration in the two types of CNS cells, neurons and oligodendrocytes, may be directly relevant to the exploration of therapeutic targets in MSA. A series of recent studies demonstrated the existence of reciprocal interactions between oligodendrocytes and neurons in neurodegeneration in MSA and in established MSA mouse models. Transgenic (Tg) mouse models in which human wild-type α-Synuclein was overexpressed in oligodendrocytes of the mouse CNS were generated under the control of different promoters [87-89]. These Tg mouse models showed abnormal aggregation of the overexpressed α-Synuclein in oligodendrocytes. In addition, the mouse lines showed that α-Synuclein accumulation in GCIs led to neuronal degeneration in the mouse CNS [88, 89]. A Tg mouse model that expressed human wild-type α-Synuclein in oligodendrocytes of the mouse CNS was generated under the control of the 2′,3′-cyclic nucleotide 3′-phosphodiesterase (CNP) promoter (CNP-Tg mouse) [88, 90]. Studies of the CNP-Tg mouse model demonstrated that the formation of oligodendrocytic α-Synuclein inclusions resulted in neuronal degeneration, as exemplified by a phenotype of motor impairment, brain atrophy that was macroscopically apparent, and a histological decrease in the number of neurons with gliosis [88]. Macroscopic brain atrophy involved the cerebrum and cerebellum. The location and severity of neuronal degeneration and astrocytosis are related to the oligodendrocytes that contained inclusions immunoreactive to α-Synuclein. The oligodendrocytes with the α-Synuclein inclusions were numerous in the cerebral white matter, particularly in the cingulate cortex and hippocampus. These oligodendrocytes and degenerated neurites were observed in the substantia nigra and inferior olivary nuclei in the brain stem and were prominent in the intermediolateral column of the spinal cord. The α-Synuclein inclusions present in oligodendrocytes induced secondary neuronal degeneration, and a similar disease process is hypothesized to underlie MSA.

Tg mouse models have provided an increased understanding of the pathophysiology of neurodegeneration in MSA, thereby clarifying the therapeutic targets of MSA. Salient characteristics relevant to MSA in humans were assessed and quantitated using the mouse models. One previous study reported that a mouse model expressing α-Synuclein in oligodendrocytes exhibited pathological features that corresponded to the motor phenotype when the Tg mice were exposed to mitochondrial inhibition by 3-nitropropionic acid [91]. The data indicated that oligodendrocytic α-Synuclein overexpression is important for an increased susceptibility to oxidative stress. This mouse model was also characterized by widespread microglial activation, and suppression of activated microglia protected nigral neurons [92].

Another study revealed that oligodendrocytic α-Synuclein inclusions induced neuronal accumulation of α-Synuclein and progressive neuronal degeneration [88]. α-Synuclein was detected in the insoluble fraction of the brain tissues of MSA patients, whereas it was distributed in the soluble fraction of the brain tissues of normal humans [88, 93]. Furthermore, α-Synuclein modulation by neuronal activity was disturbed in the mouse cells, and α-Synuclein insolubility may cause abnormal regulation of α-Synuclein synaptic accumulation, disturbing the normal function of α-Synuclein [94, 95]. Oligodendrocytic α-Synuclein accumulation induced secondary neuronal dysfunction and degeneration, and a similar disease process is hypothesized to underlie MSA. Consequently, a novel pathological mechanism of neurodegeneration was elucidated in the CNP-Tg mouse model. β-III Tubulin, a type of neuron-specific tubulin expressed in microtubules, was identified as a protein that binds to α-Synuclein in the mouse CNS [94]. Binding of this protein to α-Synuclein promoted α-Synuclein accumulation, leading to neuronal dysfunction in the CNP-Tg mouse model. The insoluble complex of α-Synuclein and β-III tubulin in microtubules accumulated progressively in neurons. This accumulation of the insoluble complex was due to the binding of α-Synuclein to β-III tubulin and not due to α-Synuclein self-aggregation [96]. Moreover, α-Synuclein accumulation was suppressed by the inhibition of microtubule polymerization. These results indicated that the binding of α-Synuclein to β-III tubulin is a key process in the development of neurodegeneration in the MSA mouse model.

Finally, a series of studies of mouse models certainly contributed to the development of a therapeutic strategy for neurodegeneration in MSA. A treatment that prevents the accumulation of insoluble α-Synuclein complexes in the mouse model has been taken forward for further testing as a potential therapeutic treatment for MSA. The data described above strongly suggest that the accumulation of insoluble α-Synuclein complexes is a therapeutic target in MSA [94]. Rifampicin, which effectively disaggregates α-Synuclein fibrils *in vitro* [97], was subsequently reported to reduce α-Synuclein in a Tg mouse model of MSA [98]. In the CNP-Tg mouse model, α-Synuclein accumulation was suppressed by treatment with a microtubule-depolymerizing agent [94, 96]. However, the results obtained after the use of the agent in the MSA mouse model raised some issues. The first issue pertained to resolving how the agent passes the blood–brain barrier and accesses the brain. Another issue pertained to avoiding or minimizing the toxicity of the microtubule-depolymerizing agent, because the agent blocks axonal transport and is potentially toxic. Future studies of MSA need to address aspects of both the pathological mechanisms of neurodegeneration and the physiological functions of α-Synuclein using mouse models and primary cultured cells, and thereby develop effective therapeutic approaches for MSA.

Cellular Mechanisms of Neurodegeneration in MSA

Tg mouse models have increased our understanding of the pathophysiology of neurodegeneration in MSA and have led to hypotheses of the involvement of three major pathological processes [69]. The first process is the increase in α-Synuclein expression in neurons via GCI formation in oligodendrocytes (Figure 2). In the normal brain, α-Synuclein is concentrated in presynaptic nerve terminals, with very little, if any, staining of this protein observed in glial cells [1]. α-Synuclein aggregates *in vitro* in a concentration-dependent manner [20], and α-Synuclein synthesis must first be upregulated to allow its accumulation to levels compatible with assembly in MSA [1]. Although the cause of α-Synuclein upregulation in oligodendrocytes remains unknown, the neuropathology of the genetic form of parkinsonism that carries the *SNCA* genomic triplication showed that α-Synuclein inclusions are found in oligodendrocytes, in addition to neuronal Lewy bodies and neuritis [44]. α-Synuclein upregulation in oligodendrocytes in the human CNS is necessary in MSA, as demonstrated experimentally via α-Synuclein overexpression selectively in oligodendrocytes in the mouse CNS [87-89]. Furthermore, the disease process includes an aberrant interaction between oligodendrocytes and neurons. Using primary cultured cells derived from the CNP-Tg mouse brain, the inhibition of glial cell proliferation by treatment with arabinofuranosyl cytidine was shown to prevent the accumulation of insoluble α-Synuclein in CNP-Tg mouse neurons [94]. Because reciprocal communication between neurons and oligodendrocytes is essential for CNS development [99], the results of studies that used the CNP-Tg mouse model suggested that degenerated oligodendrocytes regulate α-Synuclein protein expression in neurons. Moreover, the treatment of primary cultured cells derived from a non-Tg mouse brain with conditioned media derived from CNP-Tg mouse cell cultures led to the accumulation of insoluble α-Synuclein in the non-Tg mouse cells [94]. It has been proposed that an oligodendrocyte-derived signal may trigger the accumulation of insoluble α-Synuclein in neurons, resulting in the progressive neuronal degeneration that is characteristic of MSA. In those experiments, oligodendrocytic (exogenous) α-Synuclein was clearly distinguished from neuronal (endogenous) α-Synuclein. The exogenous α-Synuclein in oligodendrocytes was of human origin, whereas the endogenous α-Synuclein in neurons was of mouse origin in the mouse CNS [88, 94]. Thus, it is speculated that oligodendrocyte-derived signals may serve as a bridge between oligodendrocytic inclusions and neuronal degeneration in MSA. In another MSA mouse model, the expression of glial-derived neurotrophic factor (GDNF) was reduced, whereas infusion of GDNF ameliorated the behavioral and neuropathological deficits but had no effect on α-Synuclein expression levels [100]. Because the accumulation of insoluble α-Synuclein compromises neuronal function and viability, oligodendrocytes may release a multitude of signals that have either trophic or deleterious effects in neurons. Studies conducted using the CNP-Tg mouse model strongly suggest that oligodendrocyte-derived signals upregulate endogenous α-Synuclein in neurons, resulting in the accumulation of insoluble α-Synuclein [94]. Abnormal regulation of α-Synuclein expression by degenerated oligodendrocytes may be a trigger for the onset of neuronal degeneration in MSA.

The second process hypothesized includes neuronal α-Synuclein accumulation and binding of α-Synuclein to β-III tubulin to develop insoluble protein complexes [69]. The microtubule component β-III tubulin was identified as a protein that interacts with α-Synuclein in the CNP-Tg mouse model [94]. Microtubules are formed from the subunits of tubulin, and tubulin results in α-Synuclein accumulation *in vitro* [69, 101]. Previous histological studies have shown that GCIs in the CNS of MSA patients contained microtubules [102, 103] and were biochemically made of α-Synuclein and tubulin [104]. Studies conducted using the MSA model showed that α-Synuclein bound to β-III tubulin, which is a neuron-specific microtubule isotype of tubulin, leading to neuronal dysfunction and degeneration [88, 94]. Although the physiological role of the interaction between α-Synuclein and β-tubulin has not been completely clarified, growing evidence has demonstrated that a definite interaction exists between the two proteins. β-Tubulin was previously reported to co-immunoprecipitate with α-Synuclein in normal human brain extracts [105, 106]. Previous reports also showed that α-Synuclein interacts with heterodimeric tubulin [105], and that tubulin initiates and promotes the formation of α-Synuclein fibrils under physiological conditions *in vitro* [106]. In the MSA mouse model, administration of a microtubule-depolymerizing agent repressed the accumulation of the insoluble complex at an early stage when the treatment preceded the formation of the insoluble complex [94, 96]. Therefore, the binding of α-Synuclein to β-III tubulin and the formation of insoluble protein complexes are important processes in neurodegeneration.

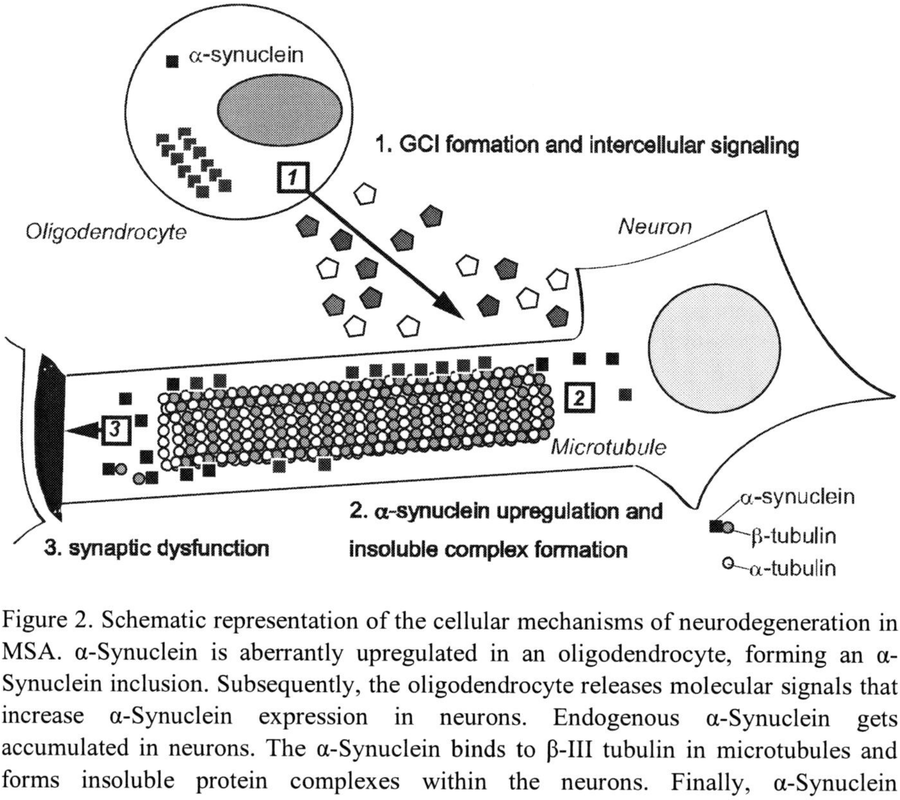

Figure 2. Schematic representation of the cellular mechanisms of neurodegeneration in MSA. α-Synuclein is aberrantly upregulated in an oligodendrocyte, forming an α-Synuclein inclusion. Subsequently, the oligodendrocyte releases molecular signals that increase α-Synuclein expression in neurons. Endogenous α-Synuclein gets accumulated in neurons. The α-Synuclein binds to β-III tubulin in microtubules and forms insoluble protein complexes within the neurons. Finally, α-Synuclein modulation by neuronal activity is disturbed, and α-Synuclein accumulation causes synaptic dysfunction, leading to neurodegeneration.

The third process hypothesized is the disturbance of α-Synuclein modulation by neuronal activity in MSA. Neuronal activity controls the synaptic accumulation of α-Synuclein, which disperses from nerve terminals in response to neuronal activity [101]. The assessment of primary cultured cells revealed that soluble α-Synuclein levels were decreased in non-Tg mouse cells in response to neuronal activity after picrotoxin (PTX) treatment, whereas α-Synuclein accumulation in response to neuronal activity after PTX treatment did not change in CNP-Tg mouse cells [94]. Because PTX is a gamma-aminobutyric acid A ($GABA_A$) receptor/chloride channel blocker, α-Synuclein regulation was disturbed in CNP-Tg mouse cells. Moreover, recent electrophysiological studies conducted in CNP-Tg mice demonstrated that the frequency of miniature inhibitory postsynaptic currents was selectively reduced [95]. A microtubule-depolymerizing agent restored the frequency of miniature inhibitory postsynaptic currents to normal levels in CNP-Tg mice. These results suggest that the α-Synuclein and β-III tubulin protein complex plays roles in the regulation of synaptic vesicle release in GABAergic interneurons by reducing inhibitory transmission [95]. The insolubility of α-Synuclein may lead to abnormal regulation of α-Synuclein synaptic accumulation, thus disturbing the normal function of α-Synuclein. Because α-Synuclein is localized to nerve terminals [6, 7, 107], when expressed at physiological levels, it functions as a negative regulator of synaptic vesicle fusion and neurotransmitter release at the synapse [108]. The total number of vesicle fusion events in α-Synuclein-overexpressing cells was significantly decreased [109]. Because the expression of endogenous mouse α-Synuclein was increased in the presynaptic terminals of CNP-Tg mouse neurons [88], the total number of synaptic vesicle fusion events may be significantly decreased, resulting in neuronal dysfunction and degeneration.

Conclusion

Synucleinopathies are a group of neurodegenerative diseases. α-Synuclein accumulation seems to be central to the pathogenesis of neurodegeneration. Although recent studies provided methods for the inhibition of α-Synuclein accumulation in synucleinopathies, future studies are necessary to elucidate the mechanisms underlying neuronal and oligodendrocytic accumulation of α-Synuclein. The development of improved experimental models of synucleinopathies will allow the identification of the cause of these diseases. Further understanding of the pathophysiology of α-Synuclein accumulation in neurodegeneration will provide new treatments for synucleinopathies.

Acknowledgments

This study was supported by JSPS KAKENHI Grant Number 25430058 and 25830045.

References

[1] Goedert, M. (2001). Alpha-synuclein and neurodegenerative diseases. *Nat Rev Neurosci*, *2*, 492-501.

[2] Ellison, D., Love, S., Chimelli, L., Harding, B. N., Lowe, J. S., Vinters, H. V., Brandner, S. & Yong, W. H. (2013). *Neuropathology: Neurodegenerative diseases* (Third edition). London, UK: Mosby.

[3] Gasser, T. (2010). α-Synuclein and Parkinson's disease. In: Schapira AHV, Lang AET, Fahn S, editors. *Movement Disorders*, *4*. Philadelphia, USA: Saunders; 40-53.

[4] Maroteaux, L., Campanelli, J. T. & Scheller, R. H. (1988). Synuclein: a neuron-specific protein localized to the nucleus and presynaptic nerve terminal. *J Neurosci*, *8*, 2804-2815.

[5] Ueda, K., Fukushima, H., Masliah, E., Xia, Y., Iwai, A., Yoshimoto, M., Otero, DA., Kondo, J., Ihara, Y. & Saitoh, T. (1993). Molecular cloning of cDNA encoding an unrecognized component of amyloid in Alzheimer disease. *Proc Natl Acad Sci USA*, *90*, 11282-11286.

[6] Jakes, R., Spillantini, M. G. & Goedert, M. (1994). Identification of two distinct synucleins from human brain. *FEBS Lett*, *345*, 27-32.

[7] Iwai, A., Masliah, E., Yoshimoto, M., Ge, N., Flanagan, L., de Silva, H. A. & Saitoh, T. (1995). The precursor protein of non-A beta component of Alzheimer's disease amyloid is a presynaptic protein of the central nervous system. *Neuron*, *14*, 467-475.

[8] Hsu, L. J., Mallory, M., Xia, Y., Veinbergs, I., Hashimoto, M., Yoshimoto, M., Thal, L. J., Saitoh, T. & Masliah, E. (1998). Expression pattern of synucleins (non-Abeta component of Alzheimer's disease amyloid precursor protein/α-Synuclein) during murine brain development. *J Neurochem*, *71*, 338-344.

[9] Schulz-Schaeffer, W. J. (2010). The synaptic pathology of alpha-synuclein aggregation in dementia with Lewy bodies, Parkinson's disease and Parkinson's disease dementia. *Acta Neuropathol (Berl)*, *120*, 131-143.

[10] Chandra, S. (2009). Synucleins. In: Squire LR, editor. *Encyclopedia of Neuroscience*. London, UK: Elsevier, *9*, 833-837.

[11] Goedert, M. & Spillantini, M. G. (2012). *Synucleinopathies and tauopathies*. In: Brady ST, Siegel GJ, Albers RW, Price DL, editors. Basic Neurochemistry. MA, USA: Academic Press; 829-843.

[12] Abeliovich, A., Schmitz, Y., Fariñas, I., Choi-Lundberg, D., Ho, W. H., Castillo, P. E., Shinsky, N., Verdugo, J. M., Armanini, M., Ryan, A.,

Hynes, M., Phillips, H., Sulzer, D. & Rosenthal, A. (2000). Mice lacking alpha-synuclein display functional deficits in the nigrostriatal dopamine system. *Neuron*, *25*, 239-252.

[13] Weinreb, P. H., Zhen, W., Poon, A. W., Conway, K. A. & Lansbury, P. T. (1996). NACP, a protein implicated in Alzheimer's disease and learning, is natively unfolded. *Biochemistry*, *35*, 13709-13715.

[14] Davidson, W. S., Jonas, A., Clayton, D. F. & George, J. M. (1998). Stabilization of α-Synuclein secondary structure upon binding to synthetic membranes. *J Biol Chem*, *273*, 9443-9449.

[15] Jo, E., McLaurin, J., Yip, C. M., St George-Hyslop, P. & Fraser, P. E. (2000). α-Synuclein membrane interactions and lipid specificity. *J Biol Chem*, *275*, 34328-34334.

[16] Perrin, R. J., Woods, W. S., Clayton, D. F. & George, J. M. (2000). Interaction of human α-Synuclein and Parkinson's disease variants with phospholipids. Structural analysis using site-directed mutagenesis. *J Biol Chem*, *275*, 34393-34398.

[17] Eliezer, D., Kutluay, E., Bussell, R. Jr. & Browne, G. (2001). Conformational properties of α-Synuclein in its free and lipid-associated states. *J Mol Biol*, *307*, 1061-1073.

[18] Conway, K. A., Harper, J. D. & Lansbury, P. T. (1998). Accelerated in vitro fibril formation by a mutant alpha-synuclein linked to early-onset Parkinson disease. *Nat Med*, *4*, 1318-1320.

[19] Narhi, L., Wood, S. J., Steavenson, S., Jiang, Y., Wu, G. M., Anafi, D., Kaufman, S. A., Martin, F., Sitney, K., Denis, P., Louis, J. C., Wypych, J., Biere, A. L. & Citron, M. (1999). Both familial Parkinson's disease mutations accelerate α-Synuclein aggregation. *J Biol Chem*, *274*, 9843-9846.

[20] Giasson, B. I., Uryu, K., Trojanowski, J. Q. & Lee, V. M. Y. (1998). Mutant and wild type human α-Synucleins assemble into elongated filaments with distinct morphologies *in vitro*. *J Biol Chem*, *274*, 7619-7622.

[21] Serpell, L. C., Berriman, J., Jakes, R., Goedert, M. & Crowther, R. A. (2000). Fiber diffraction of synthetic α-Synuclein filaments shows amyloid-like cross-beta conformation. *Proc Natl Acad Sci USA*, *97*, 4897-4902.

[22] Crowther, R. A., Jakes, R., Spillantini, M. G. & Goedert, M. (1998). Synthetic filaments assembled from C-terminally truncated alpha-synuclein. *FEBS Lett*, 436, 309-312.

[23] Fujiwara, H., Hasegawa, M., Dohmae, N., Kawashima, A., Masliah, E., Goldberg, M. S., Shen, J., Takio, K. & Iwatsubo, T. (2002). α-Synuclein is phosphorylated in synucleinopathy lesions. *Nat Cell Biol*, 4, 160-164.

[24] Ince, P. G., Clark, B., Holton, J., Revesz, T. & Wharton, S. B. (2008). Diseases of movement and system degenerations. In: Love S, Louis DN, Ellison DW, editors. *Greenfield's Neuropathology*. London: Arnold; 889-1030.

[25] Lewy, F. H. (1912). Paralysis agitans. In: Lewandowsky MH, editor. *Handbuch der Neurologie*. Berlin, Germany: Springer, *3*, 920-933.

[26] Tretiakoff, C. (1919). Contribution a l'etude de l'anatomie pathologique du Locus Niger de Soemmering avec quelques deductions relatives a la pathogenie des troubles du tonus musculaire et de la maladie de Parkinson. *Thesis*. Paris, France: Universite de Paris.

[27] Gibb, W. R. & Lees, A. J. (1988). The relevance of the Lewy body to the pathogenesis of idiopathic Parkinson's disease. *J Neurol Neurosurg Psychiatry*, *51*, 745-752.

[28] Spillantini, M. G., Schmidt, M. L., Lee, V. M. Y., Trojanowski, J. Q., Jakes, R. & Goedert, M. (1997). Alpha-synuclein in Lewy bodies. *Nature*, *388*, 839-840.

[29] Spillantini, M. G., Crowther, R. A., Jakes, R., Hasegawa, M. & Goedert, M. (1998). α-Synuclein in filamentous inclusions of Lewy bodies from Parkinson's disease and dementia with Lewy bodies. *Proc Natl Acad Sci USA*, *95*, 6469-6473.

[30] Suzuki, Y. & Yazawa, I. (2011). Pathological accumulation of atrophin-1 in dentatorubral-pallidoluysian atrophy. *Int J Clin Exp Pathol*, *4*, 378-384.

[31] Ross, C. A. & Poirier, M. A. (2004). Protein aggregation and neurodegenerative disease. *Nat Med*, *10*, S10-S17.

[32] Saudou, F., Finkbeiner, S., Devys, D. & Greenberg, M. E. (1998). Huntingtin acts in the nucleus to induce apoptosis but death does not correlate with the formation of intranuclear inclusions. *Cell*, *95*, 55-66.

[33] Kim, M., Lee, H. S., LaForet, G., McIntyre, C., Martin, E. J., Chang, P., Kim, T. W., Williams, M., Reddy, P. H., Tagle, D., Boyce, F. M., Won, L., Heller, A., Aronin, N. & DiFiglia, M. (1999). Mutant huntingtin expression in clonal striatal cells: dissociation of inclusion formation and neuronal survival by caspase inhibition. *J Neurosci*, *19*, 964-973.

[34] Arrasate, M., Mitra, S., Schweitzer, E. S., Segal, M. R. & Finkbeiner, S. (2004). Inclusion body formation reduces levels of mutant huntingtin and the risk of neuronal death. *Nature*, *431*, 805-810.

[35] Ross, C. A. (2002). Polyglutamine pathogenesis: emergence of unifying mechanisms for Huntington's disease and related disorders. *Neuron*, *35*, 819-822.

[36] Sánchez, I., Mahlke, C. & Yuan, J. (2003). Pivotal role of oligomerization in expanded polyglutamine neurodegenerative disorders. *Nature*, *421*, 373-379.

[37] Chen, L. & Feany, M. B. (2005). Alpha-synuclein phosphorylation controls neurotoxicity and inclusion formation in a *Drosophila* model of Parkinson disease. *Nat Neurosci*, 8, 657-663.

[38] Fahn, S. & Przedborski, S. (2010). *Parkinson disease*. In: Rowland LP, Pedely TA, editors. Merritt's Neurology. Twelfth edition, Philadelphia: Lippincott Williams & Wilkins; 751-769.

[39] Gelb, D. J., Oliver, E. & Gilman, S. (1999). Diagnostic criteria for Parkinson disease. *Arch Neurol*, *56*, 33-39.

[40] Braak, H., Bohl, J. R., Müller, C. M., Rüb, U., de Vos, R. A. & Del Tredici, K. (2006). The staging procedure for the inclusion body pathology associated with sporadic Parkinson's disease reconsidered. *Mov Disord*, *21*, 2042-2051.

[41] Polymeropoulos, M. H., Lavedan, C., Leroy, E., Ide, S. E., Dehejia, A., Dutra, A., Pike, B., Root, H., Rubenstein, J., Boyer, R., Stenroos, E. S., Chandrasekharappa, S., Athanassiadou, A., Papapetropoulos, T., Johnson, W. G., Lazzarini, A. M., Duvoisin, R. C., Di, Iorio, G., Golbe, L. I. & Nussbaum, R. L. (1997). Mutation in the *alpha-synuclein* gene identified in families with Parkinson's disease. *Science*, *276*, 2045-2047.

[42] Krüger, R., Kuhn, W., Müller, T., Woitalla, D., Graeber, M., Kösel, S., Przuntek, H., Epplen, J. T., Schöls, L. & Riess, O. (1998). Ala30Pro mutation in the gene encoding *alpha-synuclein* in Parkinson's disease. *Nat Genet*, *18*, 106-108.

[43] Zarranz, J. J., Alegre, J., Gómez-Esteban, J. C., Lezcano, E., Ros, R., Ampuero, I., Vidal, L., Hoenicka, J., Rodriguez, O., Atarés, B., Llorens, V., Gomez, Tortosa, E., del Ser, T., Muñoz, D. G. & de Yebenes, J. G. (2004). The new mutation, E46K, of *alpha-synuclein* causes Parkinson and Lewy body dementia. *Ann Neurol*, *55*, 164-173.

[44] Singleton, A. B., Farrer, M., Johnson, J., Singleton, A., Hague, S., Kachergus, J., Hulihan, M., Peuralinna, T., Dutra, A., Nussbaum, R., Lincoln, S., Crawley, A., Hanson, M., Maraganore, D., Adler, C., Cookson, M. R., Muenter, M., Baptista, M., Miller, D., Blancato, J., Hardy, J. & Gwinn-Hardy, K. (2003). *alpha-Synuclein* locus triplication causes Parkinson's disease. *Science*, *302*, 841.

[45] Fuchs, J., Nilsson, C., Kachergus, J., Munz, M., Larsson, E. M., Schüle, B., Langston, J. W., Middleton, F. A., Ross, O. A., Hulihan, M., Gasser, T. & Farrer, M. J. (2007). Phenotypic variation in a large Swedish pedigree due to *SNCA* duplication and triplication. *Neurology*, *68*, 916-922.

[46] Paisan-Ruíz, C., Jain, S., Evans, E. W., Gilks, W. P., Simon, J., van der Brug, M., Lopez, de Munain, A., Aparicio, S., Gil, A. M., Khan, N., Johnson, J., Martinez, J. R., Nicholl, D., Carrera, I. M., Pena, A. S., de Silva, R., Lees, A., Martí-Masso, J. F., Perez-Tur, J., Wood, N. W. & Singleton, A. B. (2004). Cloning of the gene containing mutations that cause PARK8-linked Parkinson's disease. *Neuron*, *44*, 595-600.

[47] Kitada, T., Asakawa, S., Hattori, N., Matsumine, H., Yamamura, Y., Minoshima, S., Yokochi, M., Mizuno, Y. & Shimizu, N. (1998). Mutations in the *parkin* gene cause autosomal recessive juvenile parkinsonism. *Nature*, *392*, 605-608.

[48] Valente, E. M., Abou-Sleiman, P. M., Caputo, V., Muqit, M. M., Harvey, K., Gispert, S., Ali, Z., Del Turco, D., Bentivoglio, A. R., Healy, D. G., Albanese, A., Nussbaum, R., González-Maldonado, R., Deller, T., Salvi, S., Cortelli, P., Gilks, W. P., Latchman, D. S., Harvey, R. J., Dallapiccola, B., Auburger, G. & Wood, N. W. (2004). Hereditary early-onset Parkinson's disease caused by mutations in *PINK1*. *Science*, *304*, 1158-1160.

[49] Bonifati, V., Rizzu, P., van Baren, MJ., Schaap, O., Breedveld, G. J., Krieger, E., Dekker, M. C., Squitieri, F., Ibanez, P., Joosse, M., van Dongen, J. W., Vanacore, N., van Swieten, J. C., Brice, A., Meco, G., van Duijn, C. M., Oostra, B. A. & Heutink, P. (2003). Mutations in the *DJ-1* gene associated with autosomal recessive early-onset parkinsonism. *Science*, *299*, 256-259.

[50] Hayashi, S., Wakabayashi, K., Ishikawa, A., Nagai, H., Saito, M., Maruyama, M., Takahashi, T., Ozawa, T., Tsuji, S. & Takahashi, H. (2000). An autopsy case of autosomal-recessive juvenile parkinsonism with a homozygous exon 4 deletion in the *parkin* gene. *Mov Disord*, *15*, 884-888.

[51] von Coelln, R., Thomas, B., Andrabi, S. A., Lim, K. L., Savitt, J. M., Saffary, R., Stirling, W., Bruno, K., Hess, E. J., Lee, M. K., Dawson, V. L. & Dawson, T. M. (2006). Inclusion body formation and neurodegeneration are parkin independent in a mouse model of alpha-synucleinopathy. *J Neurosci*, *26*, 3685-3696.

[52] Gupta, A. & Dawson, T. M. (2010). Pathogenesis of Parkinson's disease. In: Schapira AHV, Lang AET, Fahn S, editors. *Movement Disorders, 4.* Philadelphia, USA: Saunders; 155-169.

[53] Halliday, G. & Murphy, K. (2010). Pathology of Parkinson's disease. In: Schapira AHV, Lang AET, Fahn S, editors. *Movement Disorders, 4.* Philadelphia, USA: Saunders; 132-154.

[54] Park, J., Lee, S. B., Lee, S., Kim, Y., Song, S., Kim, S., Bae, E., Kim, J., Shong, M., Kim, J. M. & Chung, J. (2006). Mitochondrial dysfunction in *Drosophila PINK1* mutants is complemented by parkin. *Nature, 441*, 1157-1161.

[55] Clark, I. E., Dodson, M. W., Jiang, C., Cao, J. H., Huh, J. R., Seol, J. H., Yoo, S. J., Hay, B. A. & Guo, M. (2006). *Drosophila* PINK1 is required for mitochondrial function and interacts genetically with parkin. *Nature, 441*, 1162-1166.

[56] Feany, M. B. & Bender, W. W. (2000). A *Drosophila* model of Parkinson's disease. *Nature*, *404*, 394-398.

[57] Giasson, B. I., Duda, J. E., Quinn, S. M., Zhang, B., Trojanowski, J. Q. & Lee, V. M. Y. (2002). Neuronal alpha-synucleinopathy with severe movement disorder in mice expressing A53T human *alpha-synuclein. Neuron*, *34*, 521-533.

[58] Winner, B., Jappelli, R., Maji, S. K., Desplats, P. A., Boyer, L., Aigner, S., Hetzer, C., Loher, T., Vilar, M., Campioni, S., Tzitzilonis, C., Soragni, A., Jessberger, S., Mira, H., Consiglio, A., Pham, E., Masliah, E., Gage, F. H. & Riek, R. (2011). *In vivo* demonstration that alpha-synuclein oligomers are toxic. *Proc Natl Acad Sci USA*, *108*, 4194-4199.

[59] Caughey, B. (2000). Transmissible spongiform encephalopathies, amyloidoses and yeast prions: common threads? *Nat Med*, *6*, 751-754.

[60] Kane, M. D., Lipinski, W. J., Callahan, M. J., Bian, F., Durham, R. A., Schwarz, R. D., Roher, A. E. & Walker, L. C. (2000). Evidence for seeding of β-amyloid by intracerebral infusion of Alzheimer brain extracts in β-amyloid precursor protein-transgenic mice. *J Neurosci*, *20*, 3606-3611.

[61] Meyer-Luehmann, M., Coomaraswamy, J., Bolmont, T., Kaeser, S., Schaefer, C., Kilger, E., Neuenschwander, A., Abramowski, D., Frey, P., Jaton, A. L., Vigouret, J. M., Paganetti, P., Walsh, D. M., Mathews, P. M., Ghiso, J., Staufenbiel, M., Walker, L. C. & Jucker, M. (2006). Exogenous induction of cerebral β-amyloidogenesis is governed by agent and host. *Science*, *313*, 1781-1784.

[62] Clavaguera, F., Bolmont, T., Crowther, R. A., Abramowski, D., Frank, S., Probst, A., Fraser, G., Stalder, A. K., Beibel, M., Staufenbiel, M., Jucker, M., Goedert, M. & Tolnay, M. (2009). Transmission and spreading of tauopathy in transgenic mouse brain. *Nat Cell Biol*, *11*, 909-913.

[63] de Calignon, A., Polydoro, M., Suárez-Calvet, M., William, C., Adamowicz, D. H., Kopeikina, K. J., Pitstick, R., Sahara, N., Ashe, K. H., Carlson, G. A., Spires-Jones, T. L. & Hyman, B. T. (2012). Propagation of tau pathology in a model of early Alzheimer's disease. *Neuron*, *73*, 685-697.

[64] Li, J. Y., Englund, E., Holton, J. L., Soulet, D., Hagell, P., Lees, A. J., Lashley, T., Quinn, N. P., Rehncrona, S., Björklund, A., Widner, H., Revesz, T., Lindvall, O. & Brundin, P. (2008). Lewy bodies in grafted neurons in subjects with Parkinson's disease suggest host-to-graft disease propagation. *Nat Med*, *14*, 501-503.

[65] Kordower, J. H., Chu, Y., Hauser, R. A., Freeman, T. B. & Olanow, C. W. (2008). Lewy body-like pathology in long-term embryonic nigral transplants in Parkinson's disease. *Nat Med*, *14*, 504-506.

[66] Desplats, P., Lee, H. J., Bae, E. J., Patrick, C., Rockenstein, E., Crews, L., Spencer, B., Masliah, E. & Lee, S. J. (2009). Inclusion formation and neuronal cell death through neuron-to-neuron, transmission of α-Synuclein. *Proc Natl Acad Sci USA*, *106*, 13010-13015.

[67] Luk, K. C., Song, C., O'Brien, P., Stieber, A., Branch, J. R., Brunden, K. R., Trojanowski, J. Q. & Lee, V. M. Y. (2009). Exogenous α-Synuclein fibrils seed the formation of Lewy body-like intracellular inclusions in cultured cells. *Proc Natl Acad Sci USA*, *106*, 20051-20056.

[68] Luk, K. C., Kehm, V., Carroll, J., Zhang, B., O'Brien, P., Trojanowski, J. Q. & Lee, V. M. Y. (2012). Pathological α-Synuclein transmission initiates Parkinson-like neurodegeneration in nontransgenic mice. *Science*, *338*, 949-953.

[69] Yazawa, I. (2012). Tubulin: structure, function, and role in disease. In: Yamauchi W, Sokic A, editors. *Tubulin: Structure, Functions, and Roles in Disease*. Nova Science, Publishers: NY, USA; 85-100.

[70] Graham, J. G. & Oppenheimer, D. R. (1969). Orthostatic hypotension and nicotine sensitivity in a case of multiple system atrophy. *J Neurol Neurosurg Psychiatry*, *32*, 28-34.

[71] Gilman, S., Low, P. A., Quinn, N., Albanese, A., Ben-Shlomo, Y., Fowler, C. J., Kaufmann, H., Klockgether, T., Lang, A. E., Lantos, P. L., Litvan, I., Mathias, C. J., Oliver, E., Robertson, D., Schatz, I. &

Wenning, G. K. (1999). Consensus statement on the diagnosis of multiple system atrophy. *J Neurol Sci*, *163*, 94-98.

[72] Gilman, S., Wenning, G. K., Low, P. A., Brooks, D. J., Mathias, C. J., Trojanowski, J. Q., Wood, N. W., Colosimo, C., Dürr A., Fowler, C. J., Kaufmann, H., Klockgether, T., Lees, A., Poewe, W., Quinn, N., Revesz, T., Robertson, D., Sandroni, P., Seppi, K. & Vidailhet, M. (2008). Second consensus statement on the diagnosis of multiple system atrophy. *Neurology*, *71*, 670-676.

[73] Wenning, G. K., Ben-Shlomo, Y., Magalhaes, M., Daniel, S. E. & Quinn, N. P. (1994). Clinical features and natural history of multiple system atrophy. An analysis of 100 cases. *Brain*, *117*, 835-845.

[74] Hara, K., Momose, Y., Tokiguchi, S., Shimohata, M., Terajima, K., Onodera, O., Kakita, A., Yamada, M., Takahashi, H., Hirasawa, M., Mizuno, Y., Ogata, K., Goto, J., Kanazawa, I., Nishizawa, M. & Tsuji, S. (2007). Multiplex families with multiple system atrophy. *Arch Neurol*, *64*, 545-551.

[75] Esiri, M. & Perl, D. (2006). *Oppenheimer's diagnostic neuropathology*. (Third edition). London, UK: Hodder Arnold.

[76] Papp, M. I. & Lantos, P. L. (1994). The distribution of oligodendroglial inclusions in multiple system atrophy and its relevance to clinical symptomatology. *Brain*, *117*, 235-243.

[77] Ozawa, T., Paviour, D., Quinn, N. P., Josephs, K. A., Sangha, H., Kilford, L., Healy, D. G., Wood, N. W., Lees, A. J., Holton, J. L. & Revesz, T. (2004). The spectrum of pathological involvement of the striatonigral and olivopontocerebellar systems in multiple system atrophy: clinicopathological correlations. *Brain*, *127*, 2657-2671.

[78] Matsuo, A., Akiguchi, I., Lee, G. C., McGeer, E. G., McGeer, P. L. & Kimura, J. (1998). Myelin degeneration in multiple system atrophy detected by unique antibodies. *Am J Pathol*, *153*, 735-744.

[79] Jellinger, K. A. & Lantos, P. L. (2010). Papp-Lantos inclusions and the pathogenesis of multiple system atrophy: an update. *Acta Neuropathol (Berl)*, *119*, 657-667.

[80] Wenning, G. K., Stefanova, N., Jellinger, K. A., Poewe, W. & Schlossmacher, M. G. (2008). Multiple system atrophy: a primary oligodendrogliopathy. *Ann Neurol*, *64*, 239-246.

[81] Papp, M. I., Kahn, J. E. & Lantos, P. L. (1989). Glial cytoplasmic inclusions in the CNS of patients with multiple system atrophy (striatonigral degeneration, olivopontocerebellar atrophy and Shy-Drager syndrome). *J Neurol Sci*, *94*, 79-100.

[82] Arima, K., Ueda, K., Sunohara, N., Arakawa, K., Hirai, S., Nakamura, M., Tonozuka-Uehara, H. & Kawai, M. (1998). NACP/ α-Synuclein immunoreactivity in fibrillary components of neuronal and oligodendroglial cytoplasmic inclusions in the pontine nuclei in multiple system atrophy. *Acta Neuropathol (Berl)*, *96*, 439-444.

[83] Tu, P. H., Galvin, J. E., Baba, M., Giasson, B., Tomita, T., Leight, S., Nakajo, S., Iwatsubo, T., Trojanowski, J. Q. & Lee, V. M. Y. (1998). Glial cytoplasmic inclusions in white matter oligodendrocytes of multiple system atrophy brains contain insoluble α-Synuclein. *Ann Neurol*, *44*, 415-422.

[84] Spillantini, M. G., Crowther, R. A., Jakes, R., Cairns, N. J., Lantos, P. L. & Goedert, M. (1998). Filamentous α-Synuclein inclusions link multiple system atrophy with Parkinson's disease and dementia with Lewy bodies. *Neurosci Lett*, *251*, 205-208.

[85] Spillantini, M. G. & Goedert, M. (2000). The α-Synucleinopathies: Parkinson's disease, dementia with Lewy bodies, and multiple system atrophy. *Ann NY Acad Sci*, *920*, 16-27.

[86] Duda, J. E., Lee, V. M. Y. & Trojanowski, J. Q. (2000). Neuropathology of synuclein aggregates. *J Neurosci Res*, *61*, 121-127.

[87] Kahle, P. J., Neumann, M., Ozmen, L., Muller, V., Jacobsen, H., Spooren, W., Fuss, B., Mallon, B., Macklin, W. B., Fujiwara, H., Hasegawa, M., Iwatsubo, T., Kretzschmar, H. A. & Haass, C. (2002). Hyperphosphorylation and insolubility of α-Synuclein in transgenic mouse oligodendrocytes. *EMBO Rep*, *3*, 583-588.

[88] Yazawa, I., Giasson, B. I., Sasaki, R., Zhang, B., Joyce, S., Uryu, K., Trojanowski, J. Q. & Lee, V. M. Y. (2005). Mouse model of multiple system atrophy: α-Synuclein expression in oligodendrocytes causes glial and neuronal degeneration. *Neuron*, *45*, 847-859.

[89] Shults, C. W., Rockenstein, E., Crews, L., Adame, A., Mante, M., Larrea, G., Hashimoto, M., Song, D., Iwatsubo, T., Tsuboi, K. & Masliah, E. (2005). Neurological and neurodegenerative alterations in a transgenic mouse model expressing human α-Synuclein under oligodendrocyte promoter: implications for multiple system atrophy. *J Neurosci*, *25*, 10689-10699.

[90] Gravel, M., Di, Polo, A., Valera, P. B. & Braun, P. E. (1998). Four-kilobase sequence of the mouse CNP gene directs spatial and temporal expression of lacZ in transgenic mice. *J Neurosci Res*, *53*, 393-404.

[91] Stefanova, N., Reindl, M., Neumann, M., Haass, C., Poewe, W., Kahle, P. J. & Wenning, G. K. (2005). Oxidative stress in transgenic mice with

oligodendroglial alpha-synuclein overexpression replicates the characteristic neuropathology of multiple system atrophy. *Am J Pathol*, *166*, 869-876.

[92] Stefanova, N., Reindl, M., Neumann, M., Kahle, P. J., Poewe, W. & Wenning, G. K. (2007). Microglial activation mediates neurodegeneration related to oligodendroglial alpha-synucleinopathy: implications for multiple system atrophy. *Mov Disord*, *22*, 2196-2203.

[93] Dickson, D. W., Liu, W., Hardy, J., Farrer, M., Mehta, N., Uitti, R., Mark, M., Zimmerman, T., Golbe, L., Sage, J., Sima, A., D'Amato, C., Albin, R., Gilman, S. & Yen, S. H. (1999). Widespread alterations of α-Synuclein in multiple system atrophy. *Am J Pathol*, *155*, 1241-1251.

[94] Nakayama, K., Suzuki, Y. & Yazawa, I. (2009). Microtubule depolymerization suppresses α-Synuclein accumulation in a mouse model of multiple system atrophy. *Am J Pathol*, *174*, 1471-1480.

[95] Ito, H., Nakayama, K., Jin, C., Suzuki, Y. & Yazawa, I. (2012). α-Synuclein accumulation reduces GABAergic inhibitory transmission in a model of multiple system atrophy. *Biochem Biophys Res Commun*, *428*, 348-353.

[96] Nakayama, K., Suzuki, Y. & Yazawa, I. (2012). Binding of neuronal α-Synuclein to β-III tubulin and accumulation in a model of multiple system atrophy. *Biochem Biophys Res Commun*, *417*, 1170-1175.

[97] Li, J., Zhu, M., Rajamani, S., Uversky, V. N. & Fink, A. L. (2004). Rifampicin inhibits α-Synuclein fibrillation and disaggregates fibrils. *Chem Biol*, *11*, 1513-1521.

[98] Ubhi, K., Rockenstein, E., Mante, M., Patrick, C., Adame, A., Thukral, M., Shults, C. & Masliah, E. (2008). Rifampicin reduces α-Synuclein in a transgenic mouse model of multiple system atrophy. *Neuroreport*, *19*, 1271-1276.

[99] Simons, M. & Trajkovic, K. (2006). Neuron-glia communication in the control of oligodendrocyte function and myelin biogenesis. *J Cell Sci*, *119*, 4381-4389.

[100] Ubhi, K., Rockenstein, E., Mante, M., Inglis, C., Adame, A., Patrick, C., Whitney, K. & Masliah, E. (2010). Neurodegeneration in a transgenic mouse model of multiple system atrophy is associated with altered expression of oligodendroglial-derived neurotrophic factors. *J Neurosci*, *30*, 6236-6246.

[101] Alim, M. A., Hossain, M. S., Arima, K., Takeda, K., Izumiyama, Y., Nakamura, M., Kaji, H., Shinoda, T., Hisanaga, S. & Ueda, K. (2002).

Tubulin seeds α-Synuclein fibril formation. *J Biol Chem*, *277*, 2112-2117.

[102] Nakazato, Y., Yamazaki, H., Hirato, J., Ishida, Y. & Yamaguchi, H. (1990). Oligodendroglial microtubular tangles in olivopontocerebellar atrophy. *J Neuropathol Exp Neurol*, *49*, 521-530.

[103] Abe, H., Yagishita, S., Amano, N., Iwabuchi, K., Hasegawa, K. & Kowa, K. (1992). Argyrophilic glial intracytoplasmic inclusions in multiple system atrophy: immunocytochemical and ultrastructural study. *Acta Neuropathol (Berl)*, *84*, 273-277.

[104] Gai, W. P., Power, J. H., Blumbergs, P. C., Culvenor, J. G. & Jensen, P. H. (1999). α-Synuclein immunoisolation of glial inclusions from multiple system atrophy brain tissue reveals multiprotein components. *J Neurochem*, *73*, 2093-2100.

[105] Payton, J. E., Perrin, R. J., Clayton, D. F. & George, J. M. (2001). Protein-protein interactions of α-Synuclein in brain homogenates and transfected cells. *Brain Res Mol Brain Res*, *95*, 138-145.

[106] Fortin, D. L., Nemani, V. M., Voglmaier, S. M., Anthony, M. D., Ryan, T. A. & Edwards, R. H. (2005). Neural activity controls the synaptic accumulation of α-Synuclein. *J Neurosci*, *25*, 10913-10921.

[107] Withers, G. S., George, J. M., Banker, G. A. & Clayton, D. F. (1997). Delayed localization of synelfin (synuclein, NACP) to presynaptic terminals in cultured rat hippocampal neurons. *Brain Res Dev Brain Res*, *99*, 87-94.

[108] Gitler, A. D. & Shorter, J. (2007). Prime time for α-Synuclein. *J Neurosci*, *27*, 2433-2434.

[109] Larsen, K. E., Schmitz, Y., Troyer, M. D., Mosharov, E., Dietrich, P., Quazi, A. Z., Savalle, M., Nemani, V., Chaudhry, F. A., Edwards, R. H., Stefanis, L. & Sulzer, D. (2006). α-Synuclein overexpression in PC12 and chromaffin cells impairs catecholamine release by interfering with a late step in exocytosis. *J Neurosci*, *26*, 11915-11922.

In: α-Synuclein ISBN: 978-1-62948-493-8
Editors: M. Polizzi and H. C. Kanowitz

Chapter II

α-Synuclein Metabolism and Aggregation in the Pathogenesis of Parkinson's Disease

Yoshihisa Watanabe, Atsushi Tsujimura, Katsutoshi Taguchi and Masaki Tanaka *

Department of Basic Geriatrics, Graduate School of Medical Science, Kyoto Prefectural University of Medicine, Kawaramachi-Hirokoji, Kamikyo-ku, Kyoto, Japan

Abstract

α-Synuclein (α-Syn) is a major constituent of Lewy bodies, the pathological intracellular inclusion bodies of Parkinson's disease (PD), dementia with Lewy bodies, and multiple system atrophy. The intracellular aggregates or the accumulation of toxic oligomer of α-Syn is implicated in the pathogenesis of these diseases. Various forms of α-Syn, such as monomer, oligomer, and fibril, are metabolized by the ubiquitin-proteasome system, autophagy-lysosome pathway, chaperone-mediated autophagy, and other proteolytic systems. The impairment of these

* Corresponding author: Masaki Tanaka, MD, PhD, Department of Basic Geriatrics, Graduate School of Medical Science, Kyoto Prefectural University of Medicine, Kawaramachi-Hirokoji, Kamikyo-ku, Kyoto 602-8566, Japan. Tel/Fax: +81-75-251-5797; E-mail: mtanaka@koto.kpu-m.ac.jp.

systems leads to the accumulation of α-Syn, resulting in neurodegeneration. In this article, we focus on the current knowledge about the involvement of α-Syn metabolism in PD pathogenesis and discuss proteolytic systems as a therapeutic target for PD.

A. Introduction

The accumulation of misfolded/abnormal proteins, potentially toxic polypeptides, severely damages cells (Goldberg, 2003). Although misfolded proteins are generated by various physical stresses (heat, oxidation, salt, and pH) or changes in the protein properties arising from mutations, their accumulation is generally prevented by proteostasis machineries such as chaperone-mediated refolding and degradation (Figure 1). However, potentially toxic species and aggregates are intracellularly accumulated and induce cell death when proteostasis is severely compromised (e.g., aging and disease) (Hartl et al., 2011).

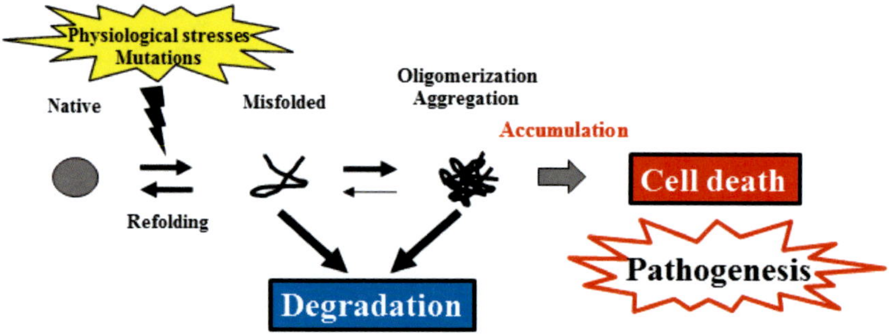

Figure 1. Protein quality control and the pathogenesis of neurodegenerative disease. A subset of cellular proteins misfolds and aggregates in times of physiological stresses (e.g., heat and oxidative conditions) or mutations. These misfolded proteins commonly undergo chaperone-mediated refolding or protein clearance by various proteolytic systems. However, cell death is promoted by the toxicity of misfolded or aggregated proteins when they are accumulated by proteostasis impairment.

The ubiquitin-proteasome system (UPS) and autophagy-lysosome pathway (ALP) are two major cellular mechanisms responsible for the clearance of misfolded and aggregate-prone proteins (Rubinsztein, 2006). Most proteasome substrates are targeted for degradation through the covalent linkage of polyubiquitin chains with a few exceptions (Murakami et al., 1992;

Watanabe & Tanaka, 2011). Ubiquitin conjugation involves ubiquitin-activating enzyme (E1), ubiquitin-conjugating enzyme (E2), and ubiquitin ligase (E3) (Ciechanover, 1994). Poly-ubiquitinated proteins are recognized by the chaperones and the ubiquitin-binding subunits of the regulatory complex of the proteasome and degraded into the catalytic core (Koga et al., 2011). In contrast to UPS, ALP was generally believed to be a non-selective bulk degradation process. However, recent evidence demonstrates that the process also be selective (Kirkin et al., 2009). This selective degradation is mediated by scaffolding adaptor proteins, such as p62/SQSTM1, NBR1, and Alfy (Kraft et al., 2010). UPS and ALP are implicated in several neurodegenerative diseases such as Parkinson's disease (PD) and in polyglutamine disorders including Huntington's disease (HD); their impairment in the brain leads to an accumulation of ubiquitin-positive inclusions (Rubinsztein, 2006).

α-Synuclein (α-Syn) is a natively unfolded protein and a major component of the intraneuronal inclusions called Lewy bodies, a pathological hallmark of PD. Three missense mutations (A30P, E46K, and A53T) and duplication and triplication of the α-Syn locus have been identified in familial PD, suggesting that conformational change and increased intracellular levels in α-Syn are implicated in PD pathogenesis (Forman et al., 2005; Savitt et al., 2006). Recent studies in animal or cultured cell models have disclosed that various proteolytic mechanisms, such as UPS, ALP, chaperone-mediated autophagy (CMA), and other proteases, contribute to α-Syn metabolism (Xilouri et al., 2013). Furthermore, it has been reported that aberrant α-Syn inhibits these proteolytic systems, resulting in further accumulation of α-Syn and neuronal degeneration.

In this article, we highlight recent insights into three themes: (B) aggregation and toxicity of α-Syn, (C) α-Syn metabolism mediated by UPS, ALP, CMA, and other proteolytic systems, and (D) PD therapeutic strategies targeted to α-Syn metabolism.

B. Structural Change in α-Synuclein (α-Syn)

α-Syn is a protein of 140 amino acids that consists of three regions: (i) the N-terminal amphipathic region (residues 1–60); (ii) the hydrophobic self-aggregating sequence known as NAC (non-Aβ component, residues 61–95); and (iii) the acidic C-terminal region (residues 96–140). Secondary structure analysis demonstrates that α-Syn is an intrinsically disordered protein under physiological conditions (Orcellet & Fernandez, 2011). A marked increase in the levels of truncated, full-length, oligomeric forms, and aggregates of α-Syn is observed in the substantia nigra (SN) of the sporadic PD brain (McNaught & Olanow, 2006). Moreover, three missense mutations (A30P, E46K, and A53T) and duplication and triplication of the α-Syn locus have been identified in familial PD, suggesting that conformational change and increased intracellular levels in α-Syn are implicated in PD pathogenesis (Forman et al., 2005; Savitt et al., 2006). α-Syn aggregation involves a multi-step nucleated polymerization process via the formation of dimmer, oligomer, protofibril, and fibril (Figure 2). Although an α-Syn fibrillar form was initially suspected as the main cause of neurodegeneration, recent evidence of potential neurotoxic quaternary structure intermediates that precede fibril formation, referred to as oligomer and protofibril, and the finding that several pathogenic mutations promote these formations suggest that the oligomer and protofibril rather than the fibrils are the pathogenic species (Volles & Lansbury, 2003; Lansbury & Lashuel, 2006). α-Syn oligomer enriched in a cross-β structure is 4 nm in height and 20 nm wide, and its molecular weight is estimated in the range of 140–800 kDa (Lundvig et al., 2005). In cell culture models, α-Syn oligomer is associated with enhanced toxicity as measured by the release of adenylate kinase from damaged cells, adenosine triphosphate levels, or caspase 3/7 activity (Kalia et al., 2013). A similar result is obtained in animal models. A transgenic fly bearing an α-Syn mutant with increased propensity to form soluble oligomers shows loss of dopaminergic neurons and motor deficits (Karpinar et al., 2009). These studies indicate that the accumulation of α-Syn toxic species is involved in PD pathogenesis, and that the prevention of this accumulation is important in PD treatment.

C. α-Syn Metabolism

C-i. Ubiquitin-Proteasome System (UPS) and Sumoylation

The UPS commonly functions in the degradation of damaged or unnecessary proteins, which are tagged with a polyubiquitin chain by ubiquitin ligases (Murata et al., 2009). The 26S proteasome is an unusually large, multi-subunit proteolytic complex, consisting of the 20S proteasome, a central catalytic machine, and 19S regulatory particles (Murata et al., 2009). The fact that Lewy bodies contain UPS-associated proteins such as ubiquitin, proteasomal subunits, and ubiquitin C-terminal hydrolase (UCH-L1) has predicted the involvement of α-Syn and a dysfunctional UPS in PD pathogenesis (Lowe et al., 1990; Kwak et al., 1991; Ii et al., 1997). α-Syn has been found to undergo ubiquitination by some E3 ubiquitin ligases such as UCH-L1, parkin, ubiquitin-protein ligase E3A (UBE3A/E6-AP), and seven in absentia homolog (SIAH). Mutations in UCH-L1 and parkin genes are also known to be responsible for familial forms of PD (Kitada et al., 1998; Leroy et al., 1998). Nonaka et al. showed that α-Syn fibril is less ubiquitinated than the soluble form and that the major ubiquitination sites of α-Syn fibril are different from those of the soluble form (Nonaka et al., 2005). In addition, other studies revealed that phosphorylation of α-Syn at Ser^{129} was required for its ubiquitination (Hasegawa et al., 2002). Despite these facts, it remains unclear whether ubiquitination is indispensable for proteasomal degradation of α-Syn. Although many studies have indicated ubiquitin-dependent proteasomal degradation of α-Syn, there is some evidence showing that α-Syn is degraded by a proteasome in a ubiquitin-independent manner. For example, different reports show that 20S and 26S proteasomes can degrade non-ubiquitinated α-Syn *in vitro* (Tofaris et al., 2001; Liu et al., 2003). Moreover, part of α-Syn purified from Lewy bodies is modified by monoubiquitination, which can serve important functions in many biological processes such as the regulation of gene transcription, protein trafficking, and DNA repair (Hasegawa et al., 2002; Tofaris et al., 2003). This evidence speculates about the involvement of α-Syn monoubiquitination in a signal for vesicular compartments trafficking because α-Syn is localized to a synaptic vesicle (Kahle et al., 2000). Similarly to ubiquitin, the small ubiquitin-like modifier (SUMO) proteins are also covalently conjugated to α-Syn (Pountney et al., 2005; Dorval & Fraser, 2006). SUMO acceptor sites in α-Syn are mapped at Lys^{96} and Lys^{102}. Intriguingly, an α-Syn mutant of sumoylation sites shows increased propensity for aggregation and cytotoxicity in a cell-based assay and increased cytotoxicity in dopaminergic neurons of the SN *in vivo*, suggesting that sumoylation of α-Syn promotes protein solubility (Krumova et al., 2011).

Although the necessity of α-Syn ubiquitination for its degradation remains obscure, treatment with a proteasome inhibitor has led to neurodegeneration and the formation of ubiquitin- and α-Syn-positive inclusions in cultured cells or primary neurons (Bennett et al., 1999; McLean et al., 2001; Rideout et al., 2004). In addition, a progressive model of PD in rats can be generated by systemic administration of PSI, a proteasome inhibitor. Postmortem studies of this rat model show neuronal loss in the SN, as well as in the locus ceruleus, nucleus basalis of Meynert, and dorsal motor nucleus of the vagus (McNaught et al., 2004). Bedford et al. established a conditional knockout mouse of proteasomal subunit PSMC1 (Rpt2), an ATPase of the 19S regulatory complex (Bedford et al., 2008). This mouse displays 26S proteasome depletion in targeted neurons (the SN or forebrain), in which the 20S proteasome is not affected. Remarkably, α-Syn- and ubiquitin-positive inclusions developed in the brain regions with impaired 26S proteasome function. In the forebrain-restricted knockout model, progressive neurodegeneration and learning deficits were observed, while in the knockout restricted to SN neurons, striking neurodegeneration and inclusion body formation occurred (Bedford et al., 2008).

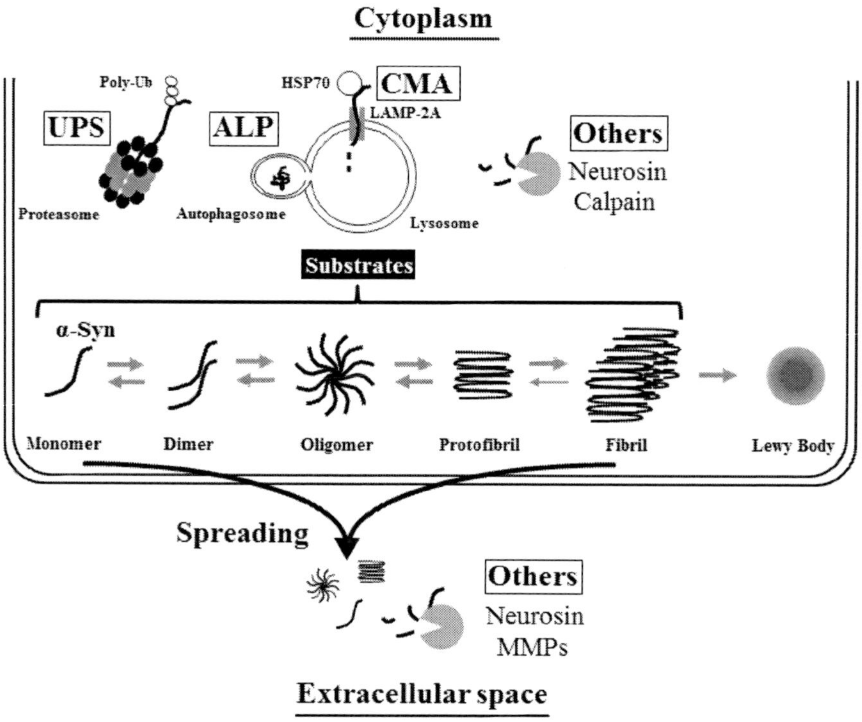

Figure 2. Structural properties and clearance mechanism of α-Syn molecular species. α-Syn undergoes structural changes in the formation process of Lewy bodies. These intermediates (dimers, oligomers, protofibrils, and fibrils) are characterized by their structures. In particular, oligomeric species take on distinct forms such as rings or stellate spheres. Various α-Syn molecular species including α-Syn monomers are degraded by UPS, ALP, CMA, and other proteases both intra- and extra-cellularly.

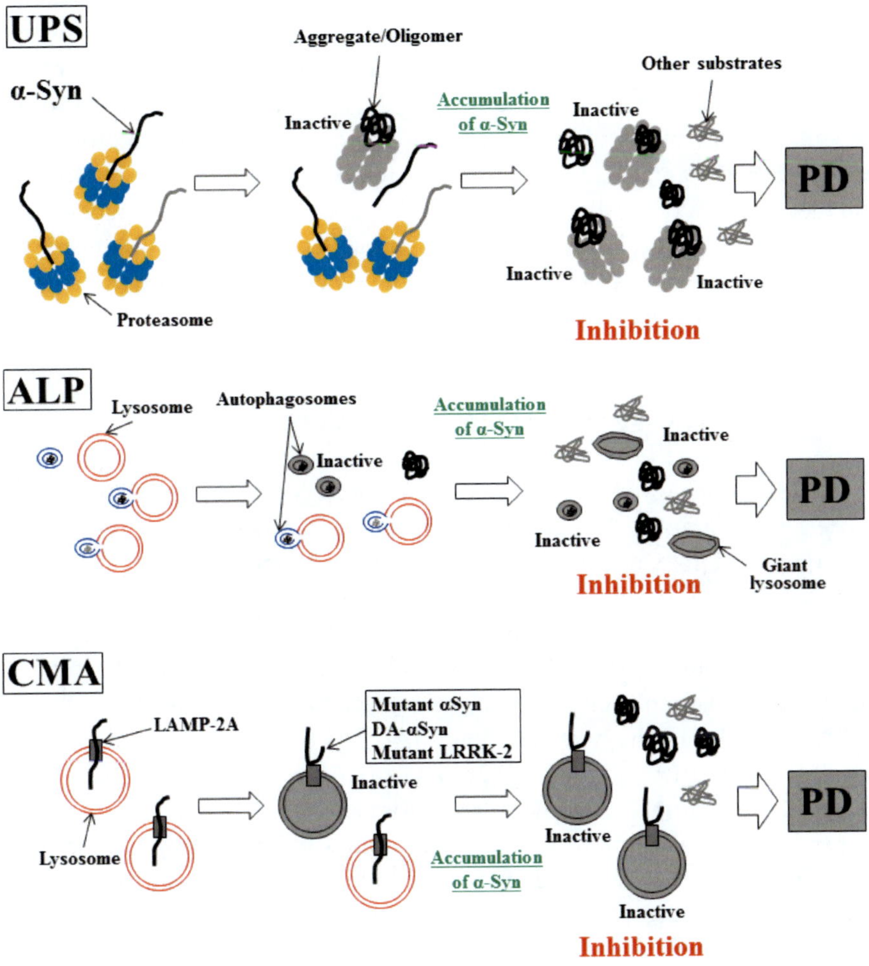

Figure 3. The models of proteostasis impairment by α-Syn aberrant forms in the pathogenesis of PD. α-Syn aberrant forms, such as oligomers, fibrils, PD mutants, and dopamine-modified forms (DA-α-Syn), inhibit several proteolytic systems. In the case of UPS, α-Syn fibril and Lewy bodies resist full entry and cause steric occlusion of the proteasome cavity, resulting in UPS dysfunction. This dysfunction leads to a further increase in α-Syn aberrant forms and neurodegeneration. Similarly, ALP is also affected by α-Syn fibrils. Although α-Syn fibrils are sequestrated into autophagosomes, they cause accumulation of immature autophagosomes. Consequently, overall macroautophagy is impaired by reducing autophagosome clearance, which may accelerate cell death. In the case of CMA, α-Syn PD mutants and DA-α-Syn inhibit its activity. In addition, it is also severely impaired by mutant proteins of another PD–associated protein, LRRK2.

What kinds of α-Syn forms are targets for proteasomal degradation? Soluble aggregation intermediates of α-Syn such as monomer, dimmer, and oligomer would be potential target substrates for proteasome because they can readily enter the 20S proteasome (Figure 2). In contrast, insoluble fibril and Lewy body may resist full entry and cause steric occlusion of the proteasome cavity. Indeed, aggregated α-Syn strongly inhibits proteasome activity *in vitro* (Snyder et al., 2003; Lindersson et al., 2004). This result suggests that attenuation of UPS activity causes the accumulation of toxic oligomers and aggregates of α-Syn, which synergistically impairs UPS proteolytic function, resulting in neurodegeneration (Figure 3).

C-ii. Autophagy-Lysosome Pathway (ALP)

Macroautophagy (hereafter called autophagy) is a highly conserved process in eukaryotes. The cytoplasm, including excess or aberrant organelles, is sequestered into double-membrane vesicles (termed an autophagosome). This autophagosome is delivered to the lysosome, for breakdown and recycling of the resulting macromolecules (Yorimitsu & Klionsky, 2005). In yeast, the isolation of autophagy-defective mutants has enabled the study of autophagy at the molecular level (Takeshige et al., 1992; Tsukada & Ohsumi, 1993), and 31 autophagy-related genes (Atgs) are identified; among them, at least 18 genes (Atg1–10, Atg12–14, Atg16–18, Atg29, and Atg31) are essential for normal autophagosome formation (Mizushima, 2007). The sequestration of substrates into autophagosomes is initiated by the formation of the phagophore (also called the isolation membrane). Various substrates are selectively/non-selectively engulfed and completely sequestrated by the elongating phagophore, and then they are delivered to the lysosome for degradation. Recent studies reveal that adaptor proteins (cargo receptors), such as p62/SQSTM1, NBR1, NDP52, and optineurin, are required for selective autophagosomal degradation (Isakson et al., 2013). Adaptor proteins are able to interact directly with both substrates and an Atg8/LC3 family member involved in autophagosome formation, resulting in a selective sequestration of substrates.

In vitro and *in vivo* studies have demonstrated that ALP is also involved in α-Syn metabolism. Autophagy inhibition with bafilomycin A1 (Baf A1) or 3-methyladenine led to an obvious increase of α-Syn levels in PC12 cells expressing α-Syn A53T mutant (Webb et al., 2003). In accordance with *in vitro* data, the accumulation of α-Syn is observed in the brain of α-Syn-EGFP transgenic mice treated with Baf A1 (Ebrahimi-Fakhari et al., 2011). Conversely, clearance of both α-Syn wild-type (WT) and mutants in cultured cells is enhanced by rapamycin treatment, which activates autophagy (Webb et al., 2003). However, it is not yet clear what kinds of α-Syn forms are the preferred substrates for ALP (Figure 2). What is known is that the exogenously introduced α-Syn fibrils can be a target for ALP. As shown in Figure 4, α-Syn fibrils are subsequently associated with p62 and LC3, which are then engulfed and digested by lysosomes when its fibrils are introduced into HEK293 cells with a transfection reagent (Watanabe et al., 2012). Conversely, direct evidence for ALP-mediated clearance of a non-ubiquitinated α-Syn monomer or oligomer has not been reported yet. However, the above-mentioned study that the α-Syn level is elevated in Baf A1-treated mice raises the possibility of these non-ubiquitinated forms as a potential target for ALP (Webb et al., 2003).

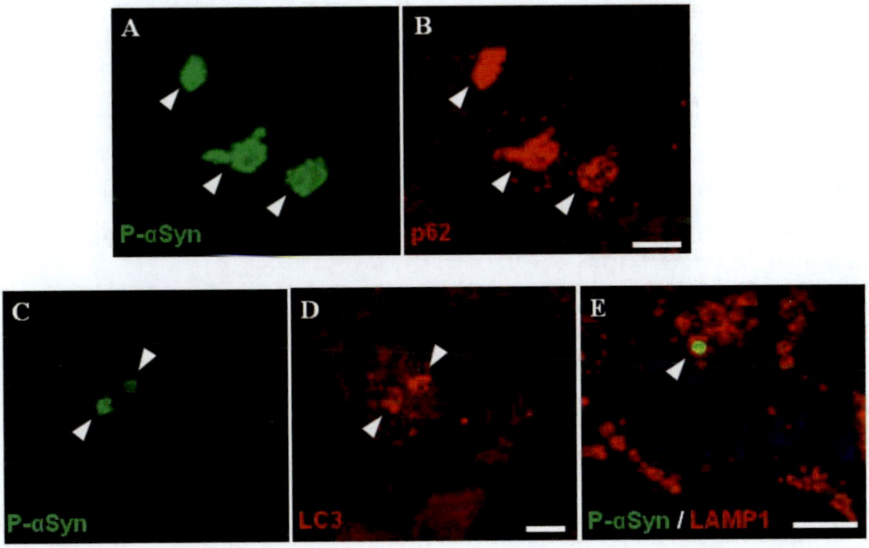

Figure 4. Clearance of α-Syn fibrils by ALP. To examine colocalization of p62 (A and B), LC3 (C and D), or LAMP-1 (E) with α-Syn fibrils, α-Syn fibrils were exogenously supplied to cultured cells, followed by immunocytochemical analysis (Watanabe et al., 2012). Phosphorylated α-Syn-positive fibrils (P-αSyn) are merged to the autophagic adaptor protein p62, the autophagosome marker LC3, or the lysosome marker LAMP-1 (arrowheads). Blue, DAPI. Scale bar, 10 μm.

It was recently demonstrated that ALP activity is impaired by Lewy body-like α-Syn inclusions. When fibrillar α-Syn is exogenously introduced into α-Syn-expressing cells, Lewy body-like inclusions are intracellularly formed (Luk et al., 2009; Volpicelli-Daley et al., 2011). As shown in Figure 3, the autophagy analysis using this culture model reveals that Lewy body-like α-Syn inclusions cause accumulation of immature autophagosomes and morphological changes in lysosomes, resulting in autophagy impairment (Tanik et al., 2013). Its severe impairment is implicated in PD pathogenesis.

C-iii. Chaperone-Mediated Autophagy (CMA)

CMA is an intracellular catabolic pathway that mediates the selective degradation of cytosolic proteins in lysosomes (Arias & Cuervo, 2011). CMA substrates possess a KFERQ-like motif that is recognized by a molecular chaperone, the heat-shock cognate protein of 70 kDa (Hsc70) (Chiang et al., 1989). Biochemical analysis shows that approximately 30% of cytosolic proteins contain this motif (Chiang & Dice, 1988). Hsc70-bound substrates are delivered to the surface of lysosomes, where they bind to a CMA receptor, the lysosome-associated membrane protein type-2A (LAMP-2A) (Cuervo & Dice, 1996). After unfolding, the substrate proteins are translocated into the lysosomal lumen in an ATP-dependent manner and finally degraded by lysosomal proteases (Arias & Cuervo, 2011).

Cuervo et al. demonstrated that α-Syn was a substrate for CMA (Cuervo et al., 2004). A potential CMA recognition motif exists in the α-Syn sequence ($_{95}$VKKDQ$_{99}$). In fact, the substitution of this motif with alanine ($_{95}$VKKAA$_{99}$) reduces the association of α-Syn with lysosomes (Cuervo et al., 2004). Interestingly, two pathogenic mutants of α-Syn (A53T and A30P) can bind to LAMP-2A but are poorly translocated into lysosomes, resulting in the disturbance of the uptake and degradation of other CMA substrates (Cuervo et al., 2004). Moreover, *in vitro* lysosomal uptake analysis reveals that dopamine-modified α-Syn (DA-α-Syn) is poorly translocated inside lysosomes and blocks degradation of other substrates by the CMA pathway, even though the binding of DA-α-Syn to lysosomes was significantly higher than that of unmodified α-Syn (Martinez-Vicente et al., 2008). This observation suggests that reduction in CMA by DA-α-Syn may contribute to selective degeneration of dopaminergic neurons in PD. Recently, it has been shown that another PD-associated protein, leucine-rich repeat kinase 2 (LRRK2), also undergoes CMA proteolysis (Orenstein et al., 2013). As is the case in α-Syn, LRRK2 PD mutants (G2019S, D1994A, and R1441C) cause a reduction in CMA activity (Orenstein et al., 2013). Furthermore, it is noteworthy that α-Syn markedly colocalizes with LAMP-2A and accumulates in neurons differentiated from induced pluripotent stem cells (iPS) of PD patients with a LRRK2-G2019S mutation (Orenstein et al., 2013). These findings predict that an α-Syn/LRRK2 self-perpetuating inhibitory effect on CMA would underlie toxicity in PD by compromising the degradation of α-Syn (Figure 3).

C-iv. Other Proteolytic Systems

Besides proteasome and autophagy, α-Syn is known to undergo proteolysis by various proteases. Intracellularly, calpain-1 and neurosin can cleave α-Syn at specific sites. Two major forms, calpain-1 and -2, are a Ca^{2+}-activated cytosolic cysteine protease and regulated by the specific endogenous inhibitor calpastatin (Wendt et al., 2004). They are involved in many processes such as skeletal remodeling, cell proliferation, differentiation, apoptosis, platelet activation, and membrane fusion (Govindarajan et al., 2008). Calpain activation seems to be implicated in PD pathogenesis. Increased expression of calpain-2 is observed in the SN and locus coeruleus of patients with PD (Mouatt-Prigent et al., 1996). Biochemical analysis demonstrated that calpains cleave α-Syn including mutant or fibril forms at multiple sites (Mishizen-Eberz et al., 2003; Mishizen-Eberz et al., 2005; Dufty et al., 2007). Interestingly, cleavage sites of α-Syn by calpain-1 are different among α-Syn forms. Calpain-1 cleaves WT α-Syn predominantly after amino acid 57 and within the NAC region (Figure 5). In contrast, it cleaves fibrillized α-Syn primarily in the region of amino acid 122 to generate fragments like those that increase susceptibility to dopamine toxicity and oxidative stress (Figure 5). Further, while calpain-1 cleaves WT α-Syn after amino acid 57, this does not occur in mutant A53T α-Syn (Figure 5). The difference in cleavage sites suggests that calpain-1-mediated cleavage of native WT α-Syn could protect against fibrillization by cleaving within the NAC region, altering the structure of this region, whereas cleavage of fibrillized α-Syn in the C-terminal region could further enhance the fibrillization process (Mishizen-Eberz et al., 2003; Mishizen-Eberz et al., 2005).

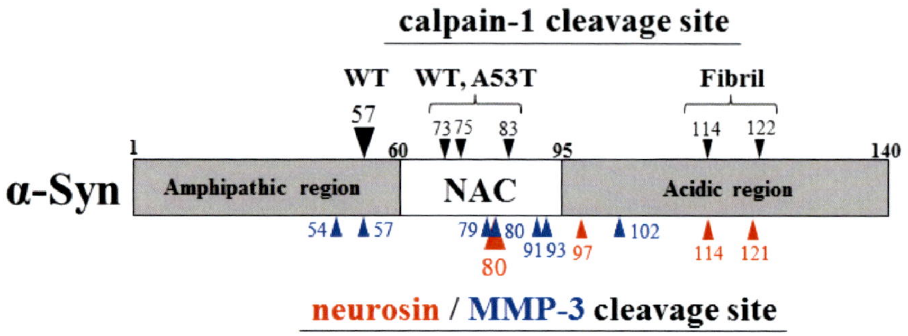

Figure 5. α-Syn cleavage sites by calpain-1, neurosin, and MMP-3 proteases. α-Syn has an N-terminal amphipathic region (1–60 amino acids), a non-amyloid-β component (NAC) domain (61–95 amino acids), and an acidic region containing several phosphorylation sites (96–140 amino acids). Its cleavage sites by each proteases are indicated by arrowheads (calpain-1; black-filled arrowheads, neurosin; red-filled arrowheads, MMP-3; blue-filled arrowheads) and numbers of amino acids. α-Syn WT-, A53T-, or fibril-specific cleavage sites by calpain-1 are also indicated at the top.

Serine protease neurosin is also involved in proteostasis of α-Syn. Neurosin belongs to the tissue kallikrein gene family and is also referred to as kallikrein 6/KLK6 (Watanabe & Tanaka, 2012). The degradation of α-Syn by this protease occurs intra- and extra-cellularly. Iwata et al. found that intracellular α-Syn is degraded by neurosin released from mitochondria into the cytosol under stressed conditions (Iwata et al., 2003). Conversely, when cultured cells expressing neurosin are incubated in a culture medium containing α-Syn, this degradation occurs in the extracellular space (Tatebe et al., 2010). Recent studies have demonstrated that α-Syn is secreted into the extracellular space via exocytosis and may move from neuron to neuron to propagate the α-Syn pathology (Lee et al., 2005; Desplats et al., 2009; Luk et al., 2012). Potential cleavage sites within α-Syn by neurosin were identified by a cleavage experiment *in vitro* (Kasai et al., 2008). Mass spectrometric analysis demonstrated that the major cleavage site is localized between Lys^{80} and Thr^{81} within the NAC region (Figure 5). In addition, the PD-associated mutant A30P and phosphorylated α-Syn are more resistant to degradation by neurosin (Kasai et al., 2008). This cleavage shows a protective effect against synucleinopathies including PD and dementia with Lewy bodies (DLB). When the neurosin gene is delivered to WT α-Syn transgenic (Tg) mice using a lentivirus vector, hippocampal neuronal loss of these mice is rescued (Spencer et al., 2013), suggesting that neurosin digestion of α-Syn reduces the generation of α-Syn toxic forms.

Furthermore, α-Syn is extracellularly cleaved by various matrix metalloproteinases (MMPs), such as MMP-1, -2, -3, and -9 (Sung et al., 2005; Levin et al., 2009). Sung et al. determined cleavage sites of *in vitro*-digested α-Syn with MMP-3 using MALDI-TOF mass spectrometry (Sung et al., 2005). As α-Syn is gradually broken down by MMP-3 from its C-terminal end, it is likely that MMP-3 would have an effect on the aggregation of α-Syn. In fact, protein aggregation of intact α-Syn is induced in the presence of the MMP-3 digested α-Syn fragments, resulting in increased cytotoxicity (Sung et al., 2005).

Taken together, some forms of α-Syn fragment cleaved by various proteases may be implicated in the pathogenesis of α-Synucleinopathies including PD, but others may contribute to a protective effect against their diseases. To achieve a clinical application in targeting proteases, it is very important to characterize individual proteolytic fragments of α-Syn.

D. Potential Targets for Parkinson's Disease (PD) Treatment

As described above, α-Syn proteostasis has been noted as a novel therapeutic target for PD. In particular, the enhancement of lysosomal function including ALP and CMA would be a promising approach. Indeed, many research groups have performed drug screening for neurodegenerative diseases such as HD and PD, as a target for these clearance systems. Rapamycin, an immunosuppressive drug, is well known to activate ALP by inhibition of the mammalian target of rapamycin (mTOR). In cultured cells, ALP activation by rapamycin increased clearance of α-Syn (Webb et al., 2003). Similarly, the same group found that trehalose, a disaccharide, also induces ALP activity via an mTOR-independent pathway, followed by accelerated α-Syn clearance (Sarkar et al., 2007a). These studies suggest that autophagy inducers would be a new approach in the treatment of PD. Novel autophagy inducers have already been identified using yeast and cultured cells. From 50,729 compounds screened, Sarkar et al. found three kinds of small molecules that activate ALP either independently or downstream of the target of rapamycin. These compounds effectively enhance the clearance of ALP substrates such as A53T α-Syn and mutant huntingtin in cultured cells (Sarkar et al., 2007b). Besides autophagy induction by these compounds, transcription factor EB (TFEB) may also be a target for therapy because TFEB overexpression stimulates autophagy and has an effect on neuroprotection via the clearance of α-Syn oligomers (Decressac et al., 2013). Similarly, CMA inducers are also identified, and the glucose-6-phosphate dehydrogenase inhibitor, 6-aminonicotinamide, and geldanamycin (an Hsp90 inhibitor) have the ability to activate CMA (Finn et al., 2005). However, whether α-Syn clearance is accelerated by these CMA-activating drugs remains to be verified

Alternatively, the direct enhancement of lysosomal function would be an effective PD treatment. Recently, other lines of evidence have also implicated the Gaucher disease (GD)-associated lysosomal enzyme glucocerebrosidase (GCase) in the pathogenesis of PD and related synucleinopathies. Indeed, mutations in GBA1, the GCase encoding gene, are known as a genetic risk factor for PD (Sidransky et al., 2009). In addition, down-regulation of GCase activity compromised lysosomal degradation and evoked α-Syn accumulation followed by neuronal toxicity in human iPS neurons derived from GD fibroblasts (Mazzulli et al., 2011). Interestingly, overexpression of α-Syn inhibits the intracellular trafficking of GCase, leading to its decreased lysosomal activity. Because of this inhibition, a pathogenic amplification loop of α-Syn accumulation and GCase dysfunction occurs, resulting in PD pathogenesis (Mazzulli et al., 2011). Thus, specific treatments that promote targeting of GCase to lysosomes are expected to diminish the formation of toxic α-Syn oligomers and break the pathogenic cycle of α-Syn aggregation and toxicity in PD.

Furthermore, the reduction of extracellular α-Syn levels is also useful in therapy for PD. As described above, the infection of a lentiviral vector expressing neurosin in the brain of α-Syn transgenic mice reduces the accumulation of α-Syn and rescues neurodegenerative alteration of its transgenic mice (Spencer et al., 2013). This may provide the way for the development of novel gene therapies for PD. Conversely, immunotherapy is also a very promising approach for the clearance of extracellular and plasma membrane-bound α-Syn. When the antibody against α-Syn is injected into the brain of α-Syn transgenic mice, antibody-assisted clearance of α-Syn occurs mainly in microglia through the Fcγ receptor (Bae et al., 2012). Another study shows that the original monoclonal antibody against the C-terminus of α-Syn is able to cross into the central nervous system and binds to cells that display α-Syn accumulation, and the antibody-antigen complex is then endocytosed and degraded by ALP (Masliah et al., 2011).

E. Perspective

We described here that various proteolytic systems are involved in α-Syn clearance, which are targets for drug development in PD and other synucleinopathies. For the establishment of the specific exclusion of pathogenic α-Syn forms, however, it is essential to identify the recognition mechanism of pathogenic α-Syn for individual proteolytic systems. Furthermore, a declining proteostasis accompanies aging and initiates an irreversible process of toxic protein aggregation and cell death (Douglas & Dillin, 2010). This implies that the reduction of α-Syn metabolism with aging contributes to PD pathogenesis. Therefore, an elucidation of the reduction mechanism of proteostasis with aging is also an important issue.

References

Arias, E. & Cuervo, A.M. (2011) Chaperone-mediated autophagy in protein quality control. *Curr Opin Cell Biol*, 23, 184-189.

Bae, E.J., Lee, H.J., Rockenstein, E., Ho, D.H., Park, E.B., Yang, N.Y., Desplats, P., Masliah, E. & Lee, S.J. (2012) Antibody-aided clearance of extracellular alpha-synuclein prevents cell-to-cell aggregate transmission. *J Neurosci*, 32, 13454-13469.

Bedford, L., Hay, D., Devoy, A., Paine, S., Powe, D.G., Seth, R., Gray, T., Topham, I., Fone, K., Rezvani, N., Mee, M., Soane, T., Layfield, R., Sheppard, P.W., Ebendal, T., Usoskin, D., Lowe, J. & Mayer, R.J. (2008) Depletion of 26S proteasomes in mouse brain neurons causes neurodegeneration and Lewy-like inclusions resembling human pale bodies. *J Neurosci*, 28, 8189-8198.

Bennett, M.C., Bishop, J.F., Leng, Y., Chock, P.B., Chase, T.N. & Mouradian, M.M. (1999) Degradation of alpha-synuclein by proteasome. *J Biol Chem*, 274, 33855-33858.

Chiang, H.L. & Dice, J.F. (1988) Peptide sequences that target proteins for enhanced degradation during serum withdrawal. *J Biol Chem*, 263, 6797-6805.

Chiang, H.L., Terlecky, S.R., Plant, C.P. & Dice, J.F. (1989) A role for a 70-kilodalton heat shock protein in lysosomal degradation of intracellular proteins. *Science*, 246, 382-385.

Ciechanover, A. (1994) The ubiquitin-proteasome proteolytic pathway. *Cell*, 79, 13-21.

Cuervo, A.M. & Dice, J.F. (1996) A receptor for the selective uptake and degradation of proteins by lysosomes. *Science*, 273, 501-503.

Cuervo, A.M., Stefanis, L., Fredenburg, R., Lansbury, P.T. & Sulzer, D. (2004) Impaired degradation of mutant alpha-synuclein by chaperone-mediated autophagy. *Science*, 305, 1292-1295.

Decressac, M., Mattsson, B., Weikop, P., Lundblad, M., Jakobsson, J. & Bjorklund, A. (2013) TFEB-mediated autophagy rescues midbrain dopamine neurons from alpha-synuclein toxicity. *Proc Natl Acad Sci U S A*, 110, E1817-1826.

Desplats, P., Lee, H.J., Bae, E.J., Patrick, C., Rockenstein, E., Crews, L., Spencer, B., Masliah, E. & Lee, S.J. (2009) Inclusion formation and neuronal cell death through neuron-to-neuron transmission of alpha-synuclein. *Proc Natl Acad Sci U S A*, 106, 13010-13015.

Dorval, V. & Fraser, P.E. (2006) Small ubiquitin-like modifier (SUMO) modification of natively unfolded proteins tau and alpha-synuclein. *J Biol Chem*, 281, 9919-9924.

Douglas, P.M. & Dillin, A. (2010) Protein homeostasis and aging in neurodegeneration. *J Cell Biol*, 190, 719-729.

Dufty, B.M., Warner, L.R., Hou, S.T., Jiang, S.X., Gomez-Isla, T., Leenhouts, K.M., Oxford, J.T., Feany, M.B., Masliah, E. & Rohn, T.T. (2007) Calpain-cleavage of alpha-synuclein: connecting proteolytic processing to disease-linked aggregation. *Am J Pathol*, 170, 1725-1738.

Ebrahimi-Fakhari, D., Cantuti-Castelvetri, I., Fan, Z., Rockenstein, E., Masliah, E., Hyman, B.T., McLean, P.J. & Unni, V.K. (2011) Distinct roles in vivo for the ubiquitin-proteasome system and the autophagy-lysosomal pathway in the degradation of alpha-synuclein. *J Neurosci*, 31, 14508-14520.

Finn, P.F., Mesires, N.T., Vine, M. & Dice, J.F. (2005) Effects of small molecules on chaperone-mediated autophagy. *Autophagy*, 1, 141-145.

Forman, M.S., Lee, V.M. & Trojanowski, J.Q. (2005) Nosology of Parkinson's disease: looking for the way out of a quagmire. *Neuron*, 47, 479-482.

Goldberg, A.L. (2003) Protein degradation and protection against misfolded or damaged proteins. *Nature*, 426, 895-899.

Govindarajan, B., Laird, J., Sherman, R., Salomon, R.G. & Bhattacharya, S.K. (2008) Neuroprotection in glaucoma using calpain-1 inhibitors: regional differences in calpain-1 activity in the trabecular meshwork, optic nerve and implications for therapeutics. *CNS Neurol Disord Drug Targets*, 7, 295-304.

Hartl, F.U., Bracher, A. & Hayer-Hartl, M. (2011) Molecular chaperones in protein folding and proteostasis. *Nature*, 475, 324-332.

Hasegawa, M., Fujiwara, H., Nonaka, T., Wakabayashi, K., Takahashi, H., Lee, V.M., Trojanowski, J.Q., Mann, D. & Iwatsubo, T. (2002) Phosphorylated alpha-synuclein is ubiquitinated in alpha-synucleinopathy lesions. *J Biol Chem*, 277, 49071-49076.

Ii, K., Ito, H., Tanaka, K. & Hirano, A. (1997) Immunocytochemical co-localization of the proteasome in ubiquitinated structures in neurodegenerative diseases and the elderly. *J Neuropathol Exp Neurol*, 56, 125-131.

Isakson, P., Holland, P. & Simonsen, A. (2013) The role of ALFY in selective autophagy. *Cell Death Differ*, 20, 12-20.

Iwata, A., Maruyama, M., Akagi, T., Hashikawa, T., Kanazawa, I., Tsuji, S. & Nukina, N. (2003) Alpha-synuclein degradation by serine protease neurosin: implication for pathogenesis of synucleinopathies. *Hum Mol Genet*, 12, 2625-2635.

Kahle, P.J., Neumann, M., Ozmen, L., Muller, V., Jacobsen, H., Schindzielorz, A., Okochi, M., Leimer, U., van Der Putten, H., Probst, A., Kremmer, E., Kretzschmar, H.A. & Haass, C. (2000) Subcellular localization of wild-type and Parkinson's disease-associated mutant alpha -synuclein in human and transgenic mouse brain. *J Neurosci*, 20, 6365-6373.

Kalia, L.V., Kalia, S.K., McLean, P.J., Lozano, A.M. & Lang, A.E. (2013) alpha-Synuclein oligomers and clinical implications for Parkinson disease. *Ann Neurol*, 73, 155-169.

Karpinar, D.P., Balija, M.B., Kugler, S., Opazo, F., Rezaei-Ghaleh, N., Wender, N., Kim, H.Y., Taschenberger, G., Falkenburger, B.H., Heise, H., Kumar, A., Riedel, D., Fichtner, L., Voigt, A., Braus, G.H., Giller, K., Becker, S., Herzig, A., Baldus, M., Jackle, H., Eimer, S., Schulz, J.B., Griesinger, C. & Zweckstetter, M. (2009) Pre-fibrillar alpha-synuclein variants with impaired beta-structure increase neurotoxicity in Parkinson's disease models. *EMBO J*, 28, 3256-3268.

Kasai, T., Tokuda, T., Yamaguchi, N., Watanabe, Y., Kametani, F., Nakagawa, M. & Mizuno, T. (2008) Cleavage of normal and pathological forms of alpha-synuclein by neurosin in vitro. *Neurosci Lett*, 436, 52-56.

Kirkin, V., McEwan, D.G., Novak, I. & Dikic, I. (2009) A role for ubiquitin in selective autophagy. *Mol Cell*, 34, 259-269.

Kitada, T., Asakawa, S., Hattori, N., Matsumine, H., Yamamura, Y., Minoshima, S., Yokochi, M., Mizuno, Y. & Shimizu, N. (1998) Mutations in the parkin gene cause autosomal recessive juvenile parkinsonism. *Nature*, 392, 605-608.

Koga, H., Kaushik, S. & Cuervo, A.M. (2011) Protein homeostasis and aging: The importance of exquisite quality control. *Ageing Res Rev*, 10, 205-215.

Kraft, C., Peter, M. & Hofmann, K. (2010) Selective autophagy: ubiquitin-mediated recognition and beyond. *Nat Cell Biol*, 12, 836-841.

Krumova, P., Meulmeester, E., Garrido, M., Tirard, M., Hsiao, H.H., Bossis, G., Urlaub, H., Zweckstetter, M., Kugler, S., Melchior, F., Bahr, M. & Weishaupt, J.H. (2011) Sumoylation inhibits alpha-synuclein aggregation and toxicity. *J Cell Biol*, 194, 49-60.

Kwak, S., Masaki, T., Ishiura, S. & Sugita, H. (1991) Multicatalytic proteinase is present in Lewy bodies and neurofibrillary tangles in diffuse Lewy body disease brains. *Neurosci Lett*, 128, 21-24.

Lansbury, P.T. & Lashuel, H.A. (2006) A century-old debate on protein aggregation and neurodegeneration enters the clinic. *Nature*, 443, 774-779.

Lee, H.J., Patel, S. & Lee, S.J. (2005) Intravesicular localization and exocytosis of alpha-synuclein and its aggregates. *J Neurosci*, 25, 6016-6024.

Leroy, E., Boyer, R., Auburger, G., Leube, B., Ulm, G., Mezey, E., Harta, G., Brownstein, M.J., Jonnalagada, S., Chernova, T., Dehejia, A., Lavedan, C., Gasser, T., Steinbach, P.J., Wilkinson, K.D. & Polymeropoulos, M.H. (1998) The ubiquitin pathway in Parkinson's disease. *Nature*, 395, 451-452.

Levin, J., Giese, A., Boetzel, K., Israel, L., Hogen, T., Nubling, G., Kretzschmar, H. & Lorenzl, S. (2009) Increased alpha-synuclein aggregation following limited cleavage by certain matrix metalloproteinases. *Exp Neurol*, 215, 201-208.

Lindersson, E., Beedholm, R., Hojrup, P., Moos, T., Gai, W., Hendil, K.B. & Jensen, P.H. (2004) Proteasomal inhibition by alpha-synuclein filaments and oligomers. *J Biol Chem*, 279, 12924-12934.

Liu, C.W., Corboy, M.J., DeMartino, G.N. & Thomas, P.J. (2003) Endoproteolytic activity of the proteasome. *Science*, 299, 408-411.

Lowe, J., McDermott, H., Landon, M., Mayer, R.J. & Wilkinson, K.D. (1990) Ubiquitin carboxyl-terminal hydrolase (PGP 9.5) is selectively present in ubiquitinated inclusion bodies characteristic of human neurodegenerative diseases. *J Pathol*, 161, 153-160.

Luk, K.C., Kehm, V., Carroll, J., Zhang, B., O'Brien, P., Trojanowski, J.Q. & Lee, V.M. (2012) Pathological alpha-synuclein transmission initiates Parkinson-like neurodegeneration in nontransgenic mice. *Science*, 338, 949-953.

Luk, K.C., Song, C., O'Brien, P., Stieber, A., Branch, J.R., Brunden, K.R., Trojanowski, J.Q. & Lee, V.M. (2009) Exogenous alpha-synuclein fibrils seed the formation of Lewy body-like intracellular inclusions in cultured cells. *Proc Natl Acad Sci U S A*, 106, 20051-20056.

Lundvig, D., Lindersson, E. & Jensen, P.H. (2005) Pathogenic effects of alpha-synuclein aggregation. *Brain Res Mol Brain Res*, 134, 3-17.

Martinez-Vicente, M., Talloczy, Z., Kaushik, S., Massey, A.C., Mazzulli, J., Mosharov, E.V., Hodara, R., Fredenburg, R., Wu, D.C., Follenzi, A., Dauer, W., Przedborski, S., Ischiropoulos, H., Lansbury, P.T., Sulzer, D. & Cuervo, A.M. (2008) Dopamine-modified alpha-synuclein blocks chaperone-mediated autophagy. *J Clin Invest*, 118, 777-788.

Masliah, E., Rockenstein, E., Mante, M., Crews, L., Spencer, B., Adame, A., Patrick, C., Trejo, M., Ubhi, K., Rohn, T.T., Mueller-Steiner, S., Seubert, P., Barbour, R., McConlogue, L., Buttini, M., Games, D. & Schenk, D. (2011) Passive immunization reduces behavioral and neuropathological deficits in an alpha-synuclein transgenic model of Lewy body disease. *PLoS One*, 6, e19338.

Mazzulli, J.R., Xu, Y.H., Sun, Y., Knight, A.L., McLean, P.J., Caldwell, G.A., Sidransky, E., Grabowski, G.A. & Krainc, D. (2011) Gaucher disease glucocerebrosidase and alpha-synuclein form a bidirectional pathogenic loop in synucleinopathies. *Cell*, 146, 37-52.

McLean, P.J., Kawamata, H. & Hyman, B.T. (2001) Alpha-synuclein-enhanced green fluorescent protein fusion proteins form proteasome sensitive inclusions in primary neurons. *Neuroscience*, 104, 901-912.

McNaught, K.S. & Olanow, C.W. (2006) Protein aggregation in the pathogenesis of familial and sporadic Parkinson's disease. *Neurobiol Aging*, 27, 530-545.

McNaught, K.S., Perl, D.P., Brownell, A.L. & Olanow, C.W. (2004) Systemic exposure to proteasome inhibitors causes a progressive model of Parkinson's disease. *Ann Neurol*, 56, 149-162.

Mishizen-Eberz, A.J., Guttmann, R.P., Giasson, B.I., Day, G.A., 3rd, Hodara, R., Ischiropoulos, H., Lee, V.M., Trojanowski, J.Q. & Lynch, D.R. (2003) Distinct cleavage patterns of normal and pathologic forms of alpha-synuclein by calpain I in vitro. *J Neurochem*, 86, 836-847.

Mishizen-Eberz, A.J., Norris, E.H., Giasson, B.I., Hodara, R., Ischiropoulos, H., Lee, V.M., Trojanowski, J.Q. & Lynch, D.R. (2005) Cleavage of alpha-synuclein by calpain: potential role in degradation of fibrillized and nitrated species of alpha-synuclein. *Biochemistry*, 44, 7818-7829.

Mizushima, N. (2007) Autophagy: process and function. *Genes Dev*, 21, 2861-2873.

Mouatt-Prigent, A., Karlsson, J.O., Agid, Y. & Hirsch, E.C. (1996) Increased M-calpain expression in the mesencephalon of patients with Parkinson's disease but not in other neurodegenerative disorders involving the mesencephalon: a role in nerve cell death? *Neuroscience*, 73, 979-987.

Murakami, Y., Matsufuji, S., Kameji, T., Hayashi, S., Igarashi, K., Tamura, T., Tanaka, K. & Ichihara, A. (1992) Ornithine decarboxylase is degraded by the 26S proteasome without ubiquitination. *Nature*, 360, 597-599.

Murata, S., Yashiroda, H. & Tanaka, K. (2009) Molecular mechanisms of proteasome assembly. *Nat Rev Mol Cell Biol*, 10, 104-115.

Nonaka, T., Iwatsubo, T. & Hasegawa, M. (2005) Ubiquitination of alpha-synuclein. *Biochemistry*, 44, 361-368.

Orcellet, M.L. & Fernandez, C.O. (2011) Structures behind the amyloid aggregation of alpha-synuclein: an NMR based approach. *Curr Protein Pept Sci*, 12, 188-204.

Orenstein, S.J., Kuo, S.H., Tasset, I., Arias, E., Koga, H., Fernandez-Carasa, I., Cortes, E., Honig, L.S., Dauer, W., Consiglio, A., Raya, A., Sulzer, D. & Cuervo, A.M. (2013) Interplay of LRRK2 with chaperone-mediated autophagy. *Nat Neurosci*, 16, 394-406.

Pountney, D.L., Chegini, F., Shen, X., Blumbergs, P.C. & Gai, W.P. (2005) SUMO-1 marks subdomains within glial cytoplasmic inclusions of multiple system atrophy. *Neurosci Lett*, 381, 74-79.

Rideout, H.J., Dietrich, P., Wang, Q., Dauer, W.T. & Stefanis, L. (2004) alpha-synuclein is required for the fibrillar nature of ubiquitinated inclusions induced by proteasomal inhibition in primary neurons. *J Biol Chem*, 279, 46915-46920.

Rubinsztein, D.C. (2006) The roles of intracellular protein-degradation pathways in neurodegeneration. *Nature*, 443, 780-786.

Sarkar, S., Davies, J.E., Huang, Z., Tunnacliffe, A. & Rubinsztein, D.C. (2007a) Trehalose, a novel mTOR-independent autophagy enhancer, accelerates the clearance of mutant huntingtin and alpha-synuclein. *J Biol Chem*, 282, 5641-5652.

Sarkar, S., Perlstein, E.O., Imarisio, S., Pineau, S., Cordenier, A., Maglathlin, R.L., Webster, J.A., Lewis, T.A., O'Kane, C.J., Schreiber, S.L. &

Rubinsztein, D.C. (2007b) Small molecules enhance autophagy and reduce toxicity in Huntington's disease models. *Nat Chem Biol*, 3, 331-338.

Savitt, J.M., Dawson, V.L. & Dawson, T.M. (2006) Diagnosis and treatment of Parkinson disease: molecules to medicine. *J Clin Invest*, 116, 1744-1754.

Sidransky, E., Nalls, M.A., Aasly, J.O., Aharon-Peretz, J., Annesi, G., Barbosa, E.R., Bar-Shira, A., Berg, D., Bras, J., Brice, A., Chen, C.M., Clark, L.N., Condroyer, C., De Marco, E.V., Durr, A., Eblan, M.J., Fahn, S., Farrer, M.J., Fung, H.C., Gan-Or, Z., Gasser, T., Gershoni-Baruch, R., Giladi, N., Griffith, A., Gurevich, T., Januario, C., Kropp, P., Lang, A.E., Lee-Chen, G.J., Lesage, S., Marder, K., Mata, I.F., Mirelman, A., Mitsui, J., Mizuta, I., Nicoletti, G., Oliveira, C., Ottman, R., Orr-Urtreger, A., Pereira, L.V., Quattrone, A., Rogaeva, E., Rolfs, A., Rosenbaum, H., Rozenberg, R., Samii, A., Samaddar, T., Schulte, C., Sharma, M., Singleton, A., Spitz, M., Tan, E.K., Tayebi, N., Toda, T., Troiano, A.R., Tsuji, S., Wittstock, M., Wolfsberg, T.G., Wu, Y.R., Zabetian, C.P., Zhao, Y. & Ziegler, S.G. (2009) Multicenter analysis of glucocerebrosidase mutations in Parkinson's disease. *N Engl J Med*, 361, 1651-1661.

Snyder, H., Mensah, K., Theisler, C., Lee, J., Matouschek, A. & Wolozin, B. (2003) Aggregated and monomeric alpha-synuclein bind to the S6' proteasomal protein and inhibit proteasomal function. *J Biol Chem*, 278, 11753-11759.

Spencer, B., Michael, S., Shen, J., Kosberg, K., Rockenstein, E., Patrick, C., Adame, A. & Masliah, E. (2013) Lentivirus mediated delivery of neurosin promotes clearance of wild-type alpha-synuclein and reduces the pathology in an alpha-synuclein model of LBD. *Mol Ther*, 21, 31-41.

Sung, J.Y., Park, S.M., Lee, C.H., Um, J.W., Lee, H.J., Kim, J., Oh, Y.J., Lee, S.T., Paik, S.R. & Chung, K.C. (2005) Proteolytic cleavage of extracellular secreted α-Synuclein via matrix metalloproteinases. *J Biol Chem*, 280, 25216-25224.

Takeshige, K., Baba, M., Tsuboi, S., Noda, T. & Ohsumi, Y. (1992) Autophagy in yeast demonstrated with proteinase-deficient mutants and conditions for its induction. *J Cell Biol*, 119, 301-311.

Tanik, S.A., Schultheiss, C.E., Volpicelli-Daley, L.A., Brunden, K.R. & Lee, V.M. (2013) Lewy body-like alpha-synuclein aggregates resist degradation and impair macroautophagy. *J Biol Chem*, 288, 15194-15210.

Tatebe, H., Watanabe, Y., Kasai, T., Mizuno, T., Nakagawa, M., Tanaka, M. & Tokuda, T. (2010) Extracellular neurosin degrades alpha-synuclein in cultured cells. *Neurosci Res*, 67, 341-346.

Tofaris, G.K., Layfield, R. & Spillantini, M.G. (2001) alpha-synuclein metabolism and aggregation is linked to ubiquitin-independent degradation by the proteasome. *FEBS Lett*, 509, 22-26.

Tofaris, G.K., Razzaq, A., Ghetti, B., Lilley, K.S. & Spillantini, M.G. (2003) Ubiquitination of alpha-synuclein in Lewy bodies is a pathological event not associated with impairment of proteasome function. *J Biol Chem*, 278, 44405-44411.

Tsukada, M. & Ohsumi, Y. (1993) Isolation and characterization of autophagy-defective mutants of Saccharomyces cerevisiae. *FEBS Lett*, 333, 169-174.

Volles, M.J. & Lansbury, P.T., Jr. (2003) Zeroing in on the pathogenic form of alpha-synuclein and its mechanism of neurotoxicity in Parkinson's disease. *Biochemistry*, 42, 7871-7878.

Volpicelli-Daley, L.A., Luk, K.C., Patel, T.P., Tanik, S.A., Riddle, D.M., Stieber, A., Meaney, D.F., Trojanowski, J.Q. & Lee, V.M. (2011) Exogenous alpha-synuclein fibrils induce Lewy body pathology leading to synaptic dysfunction and neuron death. *Neuron*, 72, 57-71.

Watanabe, Y. & Tanaka, M. (2011) p62/SQSTM1 in autophagic clearance of a non-ubiquitylated substrate. *J Cell Sci*, 124, 2692-2701.

Watanabe, Y. & Tanaka, M. (2012) *Physiological functions and pathology of neurosin/kallikrein 6 in the central nervous system*. In: Chiba, I. and Kamio, T. editors. Serine Proteases: Mechanism, Structure and Evolution. Nova Science Publishers Inc, New York. 129-137.

Watanabe, Y., Tatebe, H., Taguchi, K., Endo, Y., Tokuda, T., Mizuno, T., Nakagawa, M. & Tanaka, M. (2012) p62/SQSTM1-dependent autophagy of Lewy body-like alpha-synuclein inclusions. *PLoS One*, 7, e52868.

Webb, J.L., Ravikumar, B., Atkins, J., Skepper, J.N. & Rubinsztein, D.C. (2003) Alpha-Synuclein is degraded by both autophagy and the proteasome. *J Biol Chem*, 278, 25009-25013.

Wendt, A., Thompson, V.F. & Goll, D.E. (2004) Interaction of calpastatin with calpain: a review. *Biol Chem*, 385, 465-472.

Xilouri, M., Brekk, O.R. & Stefanis, L. (2013) Alpha-synuclein and Protein Degradation Systems: a Reciprocal Relationship. *Mol Neurobiol*, 47, 537-551.

Yorimitsu, T. & Klionsky, D.J. (2005) Autophagy: molecular machinery for self-eating. *Cell Death Differ*, 12 Suppl 2, 1542-1552.

In: α-Synuclein ISBN: 978-1-62948-493-8
Editors: M. Polizzi and H. C. Kanowitz

Chapter III

Potential Contribution of α-Synuclein Dysregulation to Parkinson's Disease Pathology

Jose L. Porras and Ruth G. Perez*
Department of Biomedical Sciences, Center of Excellence in Neurosciences, Texas Tech University Health Sciences Center at El Paso, Paul L. Foster School of Medicine, El Paso, TX, US

Abstract

The motor symptoms of Parkinson's disease (PD) arise after a marked loss of substantia nigra dopaminergic neurons. Common to the surviving nigral neurons in PD brain are proteinaceous aggregates known as Lewy bodies, which contain high levels of the presynaptic protein α-Synuclein. Despite being strongly implicated in PD, the precise role of α-Synuclein in PD neuropathology remains inconclusive. However, it is known that the α-Synuclein protein contributes to the normal functioning of dopaminergic neurons in part by interacting with a number of key regulatory proteins and other cellular components. Normally, α-Synuclein is present at low levels and in a soluble form that can interact with proteins to modulate their functions, especially those that regulate

* Corresponding Author: Ruth G. Perez, Ph.D., Associate Professor of Biomedical Sciences, 5001 El Paso Drive, MSB1 4002, El Paso, TX 79905, Phone: 915-251-4193, FAX: 915-783-1230, ruth.g.perez@ttuhsc.edu

dopamine production, release, and reuptake. In aging human brain, α-Synuclein levels rise, which triggers over-modulation of α-Synuclein interacting proteins. Eventually, high levels of α-Synuclein lead to: (i) Lewy body pathology, (ii) α-Synuclein accumulation in insoluble oligomers that impairs its normal regulatory functions, (iii) dysregulation of dopaminergic cellular function, (iv) the death of nigral dopaminergic neurons, and (v) the emergence of PD motor symptoms. This commentary will describe normal functions of α-Synuclein in dopaminergic neurons, and explore how changes in the levels and solubility of intraneuronal α-Synuclein may contribute to PD.

I. Introduction

Ever since the discovery of mutations in the α-Synuclein (a-Syn) gene that cause familial forms of Parkinson's disease (PD) [1-3], research has focused on the a-Syn protein. Furthermore, duplications and triplications of wild type human a-Syn also cause early onset PD [4, 5], although most PD is sporadic and associated with aging. Many laboratories have been exploring how a-Syn acquires a *toxic gain of function* that is especially detrimental to dopaminergic neurons of the substantia nigra pars compacta (SNc) [6-9]. Loss of nigrostriatal axons depletes striatal dopamine levels, resulting in movement problems that underlie the clinical diagnosis of PD. Translational studies in our laboratory explore normal a-Syn functions in dopaminergic neuronal cells, mouse models, and postmortem human brain to help identify therapeutic targets for PD. We believe that understanding more about normal a-Syn function will lead to therapies aimed at sustaining nigrostriatal health and SNc neuronal viability. Below, we comment on research that forms the basis of our hypothetical model in which we propose that too much or too little soluble a-Syn in dopaminergic neurons contributes to nigrostriatal damage in PD (Fig. 1).

PD is the second most common neurodegenerative disorder after Alzheimer's disease [10]. The cardinal motor symptoms of PD include bradykinesia, resting tremor, postural instability, and muscle rigidity, which arise after a marked loss of SNc dopaminergic neurons [11, 12]. Common to the surviving SNc neurons are proteinaceous cytosolic aggregates known as Lewy bodies and Lewy neurites, which contain high levels of insoluble a-Syn that is phosphorylated on serine 129 (PSer129), a modification that may accelerate a-Syn aggregation [13-15]. a-Syn is a normal presynaptic protein [16-18] that has a tendency to aggregate inside neurons in the brains of individuals with PD, Dementia with Lewy Bodies, Alzheimer's disease, and

Multiple System Atrophy, diseases collectively known as synucleinopathies [19]. The formation of Lewy bodies and Lewy neurites in SNc neurons contributes to the onset of the neuropathology underlying PD movement problems. Although a-Syn is highly implicated in PD by mutations, duplications or triplications, and its abundance in Lewy bodies and neurites, the precise contributions of the a-Syn protein to normal neuronal function and neuropathology are still emerging. Among the most intriguing discoveries about a-Syn is evidence that it can be secreted and transfer from cell to cell in a manner to stimulate pathology [20]. With regard to function, others have shown that a-Syn has homology to the 14-3-3 family of chaperone-like adapter proteins, and that 14-3-3 and a-Syn work in counterpoint to each other, to regulate the activity of key cellular proteins that they can both bind [21]. For instance, binding of 14-3-3 to phosphorylated tyrosine hydroxylase (TH; EC 1.14.16.2) stimulates TH enzymatic activity and dopamine synthesis [22-25]. We discovered that a-Syn binds TH as well as other enzymes that regulate dopamine synthesis, including aromatic amino acid decarboxylase (AADC; EC 4.1.1.28) and protein phosphatase 2A (PP2A; EC 3.1.3.16) [26-29]. We discovered that a-Syn normally localizes to mitochondria where it associates with TH [28], and further that 14-3-3ζ can stimulate dopamine synthesis on mitochondria [30]. Dopamine synthesis occurring at the mitochondrion has the potential to be problematic. Based on the data, we propose that a-Syn, 14-3-3, TH, AADC, and PP2A form a regulatory complex that acts to help optimize dopamine levels in presynaptic terminals in the striatum. In this commentary, we will review findings largely from our laboratory regarding a-Syn, TH, AADC, and PP2A and how a-Syn levels or aggregation may contribute to the pathogenesis of PD.

II. a-Synuclein

a-Syn is a small presynaptic protein [16] with chaperone-like properties [31, 32] that contributes to signal transduction by protein-protein interactions [33-38]. In addition, a-Syn interacts with membranes and vesicles [39-47] as well as with the dopamine transporter to regulate dopamine release and reuptake [48-50]. a-Syn also influences other characteristics such as membrane curvature, which may affect vesicles and other cellular organelles [51, 52]. Alterations in the levels and/or the solubility state of a-Syn could thus disrupt many processes within a dopaminergic neuron, including effects on enzymes that directly regulate dopamine biosynthesis. One such protein, TH, is further described below.

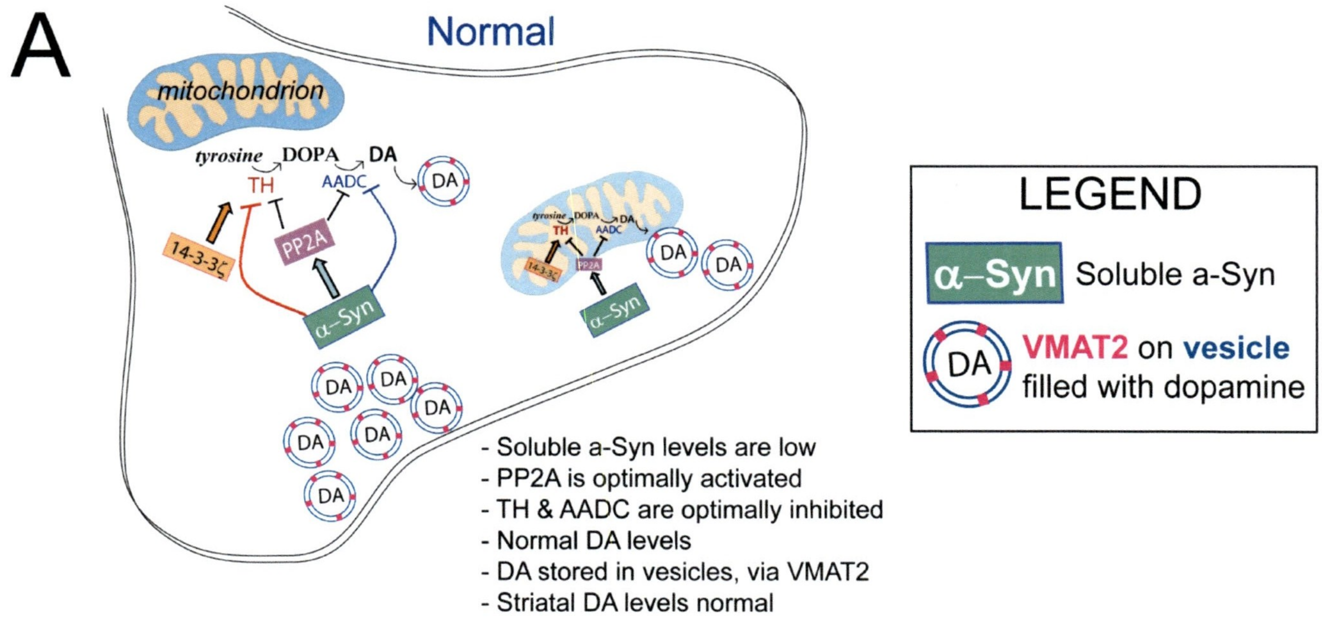
A
Normal
mitochondrion
tyrosine
DOPA
DA
TH
AADC
14-3-3ζ
PP2A
α–Syn
DA
LEGEND
α–Syn
Soluble a-Syn
VMAT2 on vesicle
filled with dopamine
- Soluble a-Syn levels are low
- PP2A is optimally activated
- TH & AADC are optimally inhibited
- Normal DA levels
- DA stored in vesicles, via VMAT2
- Striatal DA levels normal

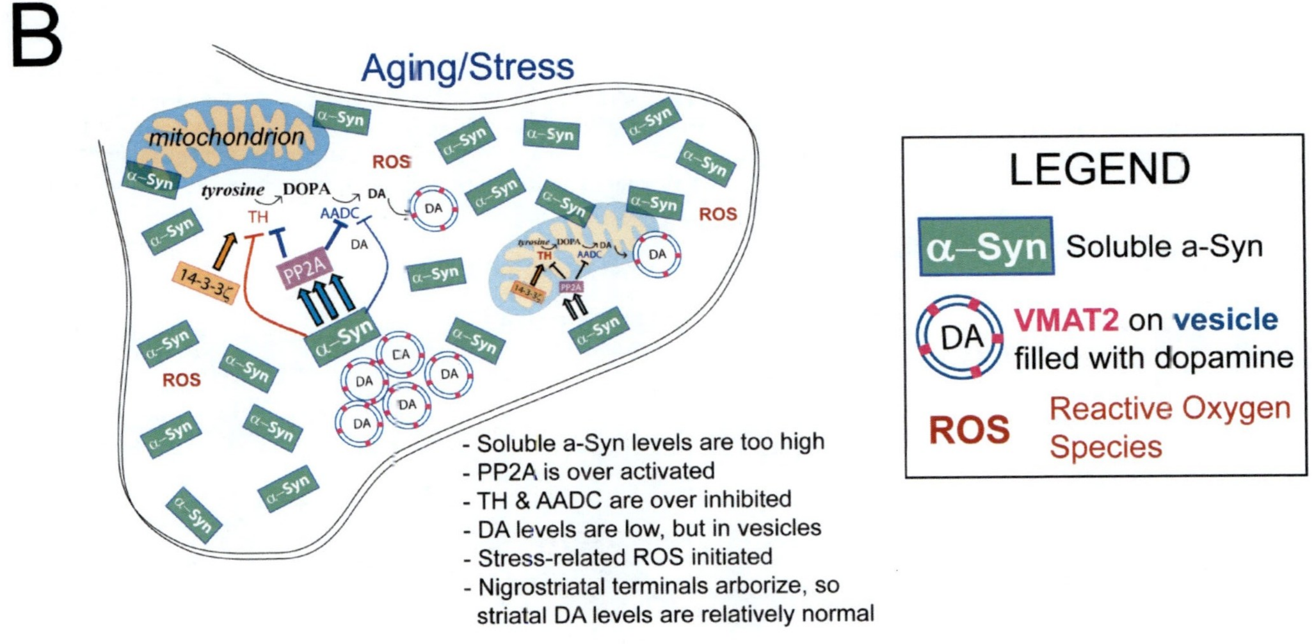

Figure 1. (Continued)

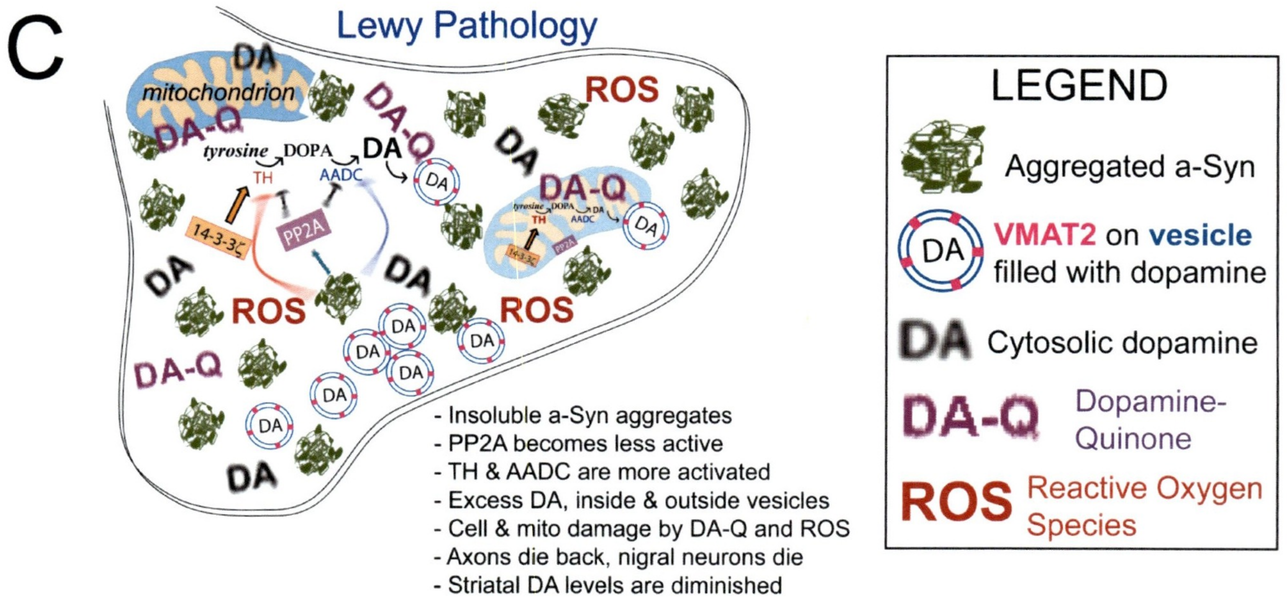

Figure 1. Hypothetical model of α-Synuclein effects on dopamine regulation in the nigrostriatal presynaptic axon terminal of normal cells, during aging and/or with chronic stress, and after soluble a-Syn loss to Lewy-like aggregation. In normal cells (**A**), a-Syn is soluble and at low levels, so it interacts with 14-3-3, TH, AADC, and PP2A in the cytoplasm and on mitochondria in a manner to optimize dopamine synthesis and dopamine packaging into presynaptic vesicles. Arrowheads in the model indicate activation; while lines with flat heads indicate inhibition.

During aging or with chronic stress (**B**), soluble a-Syn increases to high levels that over activate PP2A and down regulate TH and AADC activity, resulting in reduced amounts of presynaptic dopamine. All proteins regulated by a-Syn are affected. Associated stress begins to contribute to the generation of reactive oxygen species (ROS), which can further stress the cell. Together these factors create an environment that is conducive to Lewy body formation. Arrowheads in the model indicate activation; while lines with flat heads indicate inhibition.

Ultimately, (**C**) a-Syn becomes insoluble and forms a Lewy neurite, causing soluble a-Syn levels to become too low to stimulate PP2A activity or to inhibit TH and AADC. This causes dopamine (DA) levels to become so high that they exceed the capacity of vesicle storage by VMAT2, resulting in free cytosolic dopamine. Unpackaged dopamine produces reactive oxygen species and highly reactive dopamine quinones (DA-Q) that can damage proteins, lipids, and DNA. Because dopamine is synthesized at the mitochondrion, dopamine related damage impairs mitochondrial health. Cumulative damage ultimately kills the nigrostriatal dopaminergic neurons. Arrowheads in the model indicate activation; while lines with flat heads indicate inhibition.

III. Tyrosine Hydroxylase

TH is the rate limiting enzyme in catecholamine biosynthesis, which acts by converting the amino acid tyrosine to 3,4-dihydroxyphenylalanine (L-DOPA). In the developing human SNc, a-Syn localizes to dopaminergic TH neurons where a-Syn plays a role in their maturation [53]. a-Syn interacts with TH in a manner to inhibit TH activity in vitro and in vivo [26-28]. TH enzymatic activity is normally stimulated by phosphorylation of key serine residues in the N-terminal regulatory domain [54]. Dopaminergic cell lines that overexpress soluble a-Syn exhibit significantly reduced TH phosphorylation compared to control cells [26-28]. Moreover, both wild type and A53T mutant human a-Syn overexpression in mice effectively reduces phosphoserine levels on TH, leading to reduced TH activity without altering TH protein levels [26]. Although a-Syn mutations (A53T, A30P, and E46K) spur Lewy body pathology, they have little effect on a-Syn functions in dopaminergic cells [7]. However, additional data reveal that too much a-Syn, at levels that are high enough to stimulate a-Syn aggregation, produces a loss of soluble a-Syn function that allows hyperphosphorylation of TH [55]. Hyperphosphorylation of TH increases dopamine synthesis, as shown by silencing a-Syn expression in MN9D dopaminergic cells, which increases TH phosphorylation, TH activity, as well as dopamine synthesis [56]. Interestingly, phosphorylation of a-Syn Ser129 can attenuate the ability of a-Syn to inhibit TH in vitro and in vivo [26], while also contributing to Lewy body/Lewy neurite formation [13-15].

IV. Aromatic Amino Acid Decarboxylase

The conversion of L-DOPA to dopamine is normally mediated by the next enzyme after TH in the catecholamine biosynthetic pathway, AADC [57]. In cells overexpressing wild type or A53T mutant human a-Syn, administration of L-DOPA was expected to stimulate dopamine synthesis by bypassing the TH rate-limiting step; however, a-Syn cells continued to have diminished dopamine synthesis even after high level L-DOPA treatment [29]. This occurs in part by a direct binding of a-Syn to AADC, which produces a parallel reduction in AADC phosphorylation that lowers AADC enzymatic activity [29, 58]. AADC in a-Syn overexpressing cells is 65% less phosphorylated than AADC in plasmid transfected control cells that express low endogenous levels of soluble a-Syn [29]. Similar to what was found for TH, overexpressing a-Syn to levels that induce a-Syn aggregation has the potential to impair a-Syn-mediated AADC inhibition, producing a subsequent increase in dopamine synthesis [7, 55]. Both TH and AADC activation appear to be regulated by increasing their phosphorylation state, which can be reversed by PP2A-mediated dephosphorylation [26, 27, 58, 59]. Even though direct binding can occur between a-Syn, TH, and AADC, we think that stimulation of PP2A activity by a-Syn likely plays the major role in dopamine regulation.

V. Protein Phosphatase 2A

PP2A is a major Ser/Thr phosphatase that contributes widely to normal brain function, including dopamine production [60]. The PP2A holoenzyme consists of three independent subunits, a catalytic C subunit, a scaffolding A subunit, and several different regulatory/targeting B subunits that help localize PP2A to specific cellular microdomains, providing functional diversity. By co-immunoprecipitation studies we have shown that a-Syn normally binds to and stimulates the activity of the C subunit of PP2A in rat brain and dopaminergic PC12 cells as well as in MN9D cells, particularly the pool of PP2A that regulates dopamine synthesis [27]. This appears to be a normal regulatory function of a-Syn, contributing to optimal PP2A activity to regulate the phosphorylation state of other a-Syn interacting proteins. In the same cells, phosphorylation of Akt, another key PP2A substrate [29] is unaltered by a-Syn overexpression; demonstrating that a-Syn interacts with unique pools of PP2A within dopaminergic cells. PP2A activity can be overstimulated by high levels of soluble a-Syn, causing dephosphorylation of AADC and TH and consequently reducing dopamine levels [26, 29]. Aside from driving Lewy body formation, a-Syn PSer129 phosphorylation lessens the stimulatory effect of a-Syn on PP2A activation [26]. We and others find that PP2A does not dephosphorylate a-Syn PSer129 [26, 61], while another report finds that certain PP2A holoenzymes can do so [62]. Regardless, postmortem brains from individuals with Dementia with Lewy Bodies or the a-Syn triplication, which contain high levels of aggregated a-Syn with high PSer129 levels, have impaired PP2A activity [63] as do Alzheimer brains [64], which often have coincident synucleinopathy. Cumulatively, the data suggest that maintaining low a-Syn levels in soluble form will help sustain optimal PP2A activity. We further propose that treatments that stimulate PP2A activity may be beneficial for the treatment of PD.

Conclusion and Hypothetical Model

In addition to a-Syn and other proteins [65], TH and 14-3-3 have also been identified in Lewy bodies [66-69]. In contrast to a-Syn, which is normally expressed at low levels [28], TH and 14-3-3 are highly expressed in dopaminergic neurons [70, 71] and there is no evidence that levels of TH or 14-3-3 decrease in neurons with Lewy pathology. Normal a-Syn function in dopaminergic neurons is controlled, at least in part, by a-Syn levels and solubility. Here we propose a model, based largely on work from our laboratory, characterizing how a-Syn contributes to both wellness and disease. In normal healthy neurons, *soluble a-Syn is present at low levels* and works in counterpoint to 14-3-3ζ to help optimize dopamine levels (Fig. 1A) [7, 21, 28, 30]. With aging and stress, soluble a-Syn levels begin to rise and *a-Syn overexpression produces too great an effect on key interacting proteins,* contributing to a down regulation in dopamine synthesis in SNc presynaptic axonal terminals in the striatum (Fig 1B) [72, 73]. As this occurs, nigrostriatal axon terminals arborize and/or up regulate TH levels, to help sustain striatal dopamine levels [11, 74, 75]. The high levels of a-Syn provide a pool that not only stimulates oligomer formation [76-80] but also kindles neuroinflammation [37, 81-84]. When a-Syn aggregates, *a loss of a-Syn regulatory function occurs in dopaminergic neurons* (Fig. 1C) [85], similar to what we noted in a-Syn knockout mice [26]. Insoluble a-Syn cannot optimally stimulate PP2A activity [63] leading to hyperphosphorylation/hyperactivity of TH and AADC [55] which can lead to an overproduction of dopamine. Cytosolic dopamine levels rise when neurotransmitter levels exceed the capacity of the vesicular monoamine transporter (VMAT2) to package dopamine in vesicles [86, 87]. Dopamine that is free in the cytosol contributes to the generation of reactive oxygen species (ROS) and dopamine quinones (DA-Q), either spontaneously or through transition metal ions and enzyme catalyzed reactions [88-95]. ROS and DA-Q that are not neutralized can injure neurons by damaging proteins, membranes, lipids, mitochondria, and DNA. This compromises the health of the nigrostriatal presynaptic terminals in striatum, causing them to die back, and eventually inducing SNc neuronal death to allow PD motor symptoms to emerge.

The staple PD treatment, L-DOPA (levodopa), can effectively restore striatal dopamine levels to help improve movement in the short run [96], but eventually the treatment contributes to motor complications such as dyskinesia [97]. Unfortunately, PD continues to progress in individuals treated with drugs that address the motor symptoms, making the search for neuroprotective strategies an important area of PD research [98]. Understanding the normal function(s) of a-Syn in dopaminergic neurons thus holds promise for identifying therapeutic targets for PD. This becomes all the more important as accumulating data show that the incidence of PD is increasing worldwide [99] and further, that a-Syn aggregation also contributes to early prodromal symptoms of PD such as diminished gastric motility and olfactory impairment [100, 101]. Synucleinopathy assessment before motor onset may thus be a valuable biomarker for early PD detection [102]. Patients could further benefit by early detection using a-Syn neuroimaging compounds that are currently in development. Such analyses could help identify those with more intact nigrostriatal pathways who may respond well to emerging neuroprotective therapies aimed at slowing or reversing progressive PD pathology. Discovering safe remedies that can help sustain low a-Syn levels, reduce neuroinflammation, and normalize PP2A activity may even act to prevent PD onset at some time in the not too distant future.

Acknowledgements

We thank M. Chattopadhyay, W. Godfrey, S. Gonzalez, S. Krishnamachari, J. Vargas, G. Vidal, and E. Villanueva for editorial advice, and dedicate this to M.J. Fox, R. Byer, J. Cordy, and in memory of L. "Rusty" Lanelli. Funding was generously provided by the Scaife Family Foundation, Michael J. Fox Foundation, and the National Institutes of Health [NINDS/R01NS042094, RGP].

References

[1] Kruger, R., et al., Ala30Pro mutation in the gene encoding alpha-synuclein in Parkinson's disease. *Nat Genet,* 1998. 18(2): p. 106-8.

[2] Polymeropoulos, M.H., et al., Mutation in the alpha-synuclein gene identified in families with Parkinson's disease. *Science*, 1997. 276(5321): p. 2045-7.

[3] Zarranz, J., et al., The new mutation, E46K, of alpha-synuclein causes parkinson and Lewy body dementia. *Ann Neurol,* 2004. 55(2): p. 164-73.

[4] Singleton, A.B., et al., alpha-Synuclein locus triplication causes Parkinson's disease. *Science*, 2003. 302(5646): p. 841.
[5] Chartier-Harlin, M.-C., et al., alpha-Synuclein locus duplication as a cause of familial Parkinson's disease. *Lancet*, 2004. 364(9440): p. 1167-1169.
[6] Goedert, M., Alpha-synuclein and neurodegenerative diseases. *Nat Rev Neurosci,* 2001. 2(7): p. 492-501.
[7] Perez, R.G. and T.G. Hastings, Could a loss of alpha-synuclein function put dopaminergic neurons at risk? *J Neurochem,* 2004. 89(6): p. 1318-24.
[8] Venda, L.L., et al., alpha-Synuclein and dopamine at the crossroads of Parkinson's disease. *Trends Neurosci,* 2010. 33(12): p. 559-68.
[9] Zhu, Y., J. Zhang, and Y. Zeng, Overview of tyrosine hydroxylase in Parkinson's disease. *CNS Neurol Disord Drug Targets,* 2012. 11(4): p. 350-8.
[10] Braak, H. and E. Braak, Pathoanatomy of Parkinson's disease. *J Neurol,* 2000. 247 Suppl 2: p. II3-10.
[11] Bernheimer, H., et al., Brain dopamine and the syndromes of Parkinson and Huntington. Clinical, morphological and neurochemical correlations. *J Neurol Sci,* 1973. 20(4): p. 415-55.
[12] Rao, G., et al., Does this patient have Parkinson disease? *JAMA*, 2003. 289(3): p. 347-53.
[13] Anderson, J.P., et al., Phosphorylation of Ser-129 is the dominant pathological modification of alpha-synuclein in familial and sporadic Lewy body disease. *J Biol Chem*, 2006. 281(40): p. 29739-52.
[14] Fujiwara, H., et al., alpha-Synuclein is phosphorylated in synucleinopathy lesions. *Nat Cell Biol,* 2002. 4(2): p. 160-4.
[15] Spillantini, M.G., et al., alpha-Synuclein in filamentous inclusions of Lewy bodies from Parkinson's disease and dementia with lewy bodies. *Proc Natl Acad Sci U S A*, 1998. 95(11): p. 6469-73.
[16] Maroteaux, L., J.T. Campanelli, and R.H. Scheller, Synuclein: a neuron-specific protein localized to the nucleus and presynaptic nerve terminal. *J Neurosci*, 1988. 8(8): p. 2804-15.
[17] Clayton, D.F. and J.M. George, The synucleins: a family of proteins involved in synaptic function, plasticity, neurodegeneration and disease. *Trends Neurosci,* 1998. 21(6): p. 249-54.
[18] Clayton, D.F. and J.M. George, Synucleins in synaptic plasticity and neurodegenerative disorders. *J Neurosci Res*, 1999. 58(1): p. 120-9.

[19] Galvin, J.E., V.M. Lee, and J.Q. Trojanowski, Synucleinopathies: clinical and pathological implications. *Arch Neurol*, 2001. 58(2): p. 186-90.

[20] Steiner, J.A., E. Angot, and P. Brundin, A deadly spread: cellular mechanisms of alpha-synuclein transfer. *Cell Death Differ*, 2011. 18(9): p. 1425-33.

[21] Ostrerova, N., et al., alpha-Synuclein shares physical and functional homology with 14-3-3 proteins. *J Neurosci,* 1999. 19(14): p. 5782-91.

[22] Toska, K., et al., Phosphorylation of tyrosine hydroxylase in isolated mice adrenal glands. *Ann N Y Acad Sci*, 2002. 971: p. 66-8.

[23] Dunkley, P.R., et al., Tyrosine hydroxylase phosphorylation: regulation and consequences. *J Neurochem*, 2004. 91(5): p. 1025-43.

[24] Ichimura, T., et al., Molecular cloning of cDNA coding for brain-specific 14-3-3 protein, a protein kinase-dependent activator of tyrosine and tryptophan hydroxylases. *Proc Natl Acad Sci USA,* 1988. 85(19): p. 7084-8.

[25] Tanji, M., et al., Activation of protein kinase C by purified bovine brain 14-3-3: comparison with tyrosine hydroxylase activation. *J Neurochem,* 1994. 63(5): p. 1908-16.

[26] Lou, H., et al., Serine 129 phosphorylation reduces the ability of alpha-synuclein to regulate tyrosine hydroxylase and protein phosphatase 2A in vitro and in vivo. *J Biol Chem,* 2010. 285(23): p. 17648-61.

[27] Peng, X., et al., Alpha-synuclein activation of protein phosphatase 2A reduces tyrosine hydroxylase phosphorylation in dopaminergic cells. *J Cell Sci,* 2005. 118(Pt 15): p. 3523-30.

[28] Perez, R.G., et al., A role for alpha-synuclein in the regulation of dopamine biosynthesis. *J Neurosci,* 2002. 22(8): p. 3090-9.

[29] Tehranian, R., et al., Alpha-synuclein inhibits aromatic amino acid decarboxylase activity in dopaminergic cells. *J Neurochem,* 2006. 99(4): p. 1188-96.

[30] Wang, J., et al., 14-3-3ζ contributes to tyrosine hydroxylase activity in MN9D cells: localization of dopamine regulatory proteins to mitochondria. *J Biol Chem,* 2009. 284(21): p. 14011-9.

[31] Souza, J.M., et al., Chaperone-like activity of synucleins. *FEBS Lett,* 2000. 474(1): p. 116-9.

[32] Kim, T.D., et al., Structural changes in alpha-synuclein affect its chaperone-like activity in vitro. *Protein Sci,* 2000. 9(12): p. 2489-96.

[33] Aoki, R. and Y.R. Li, alpha-Synuclein promotes neuroprotection through NF-kappaB-mediated transcriptional regulation of protein kinase Cdelta. *Sci Signal,* 2011. 4(195): p. jc6.

[34] Hashimoto, M., et al., alpha-Synuclein protects against oxidative stress via inactivation of the c-Jun N-terminal kinase stress-signaling pathway in neuronal cells. *J Biol Chem*, 2002. 277(13): p. 11465-72.

[35] Jensen, P.H., et al., alpha-synuclein binds to Tau and stimulates the protein kinase A-catalyzed tau phosphorylation of serine residues 262 and 356. *J Biol Chem,* 1999. 274(36): p. 25481-9.

[36] Jin, H., et al., alpha-Synuclein negatively regulates protein kinase Cdelta expression to suppress apoptosis in dopaminergic neurons by reducing p300 histone acetyltransferase activity. *J Neurosci,* 2011. 31(6): p. 2035-51.

[37] Klegeris, A., et al., Alpha-synuclein activates stress signaling protein kinases in THP-1 cells and microglia. *Neurobiol Aging,* 2008. 29(5): p. 739-52.

[38] Seo, J.H., et al., Alpha-synuclein regulates neuronal survival via Bcl-2 family expression and PI3/Akt kinase pathway. *Faseb J,* 2002. 16(13): p. 1826-8.

[39] Maroteaux, L. and R.H. Scheller, The rat brain synucleins; family of proteins transiently associated with neuronal membrane. *Brain Res Mol Brain Res,* 1991. 11(3-4): p. 335-43.

[40] Davidson, W.S., et al., Stabilization of alpha-synuclein secondary structure upon binding to synthetic membranes. *J Biol Chem,* 1998. 273(16): p. 9443-9.

[41] Narayanan, V. and S. Scarlata, Membrane binding and self-association of alpha-synucleins. *Biochemistry*, 2001. 40(33): p. 9927-34.

[42] Bisaglia, M., et al., The 11-mer repeats of human alpha-synuclein in vesicle interactions and lipid composition discrimination: a cooperative role. *Biopolymers*, 2006. 84(3): p. 310-6.

[43] Bartels, T., et al., The N-terminus of the intrinsically disordered protein alpha-synuclein triggers membrane binding and helix folding. *Biophys J,* 2010. 99(7): p. 2116-24.

[44] Varkey, J., et al., Membrane curvature induction and tubulation are common features of synucleins and apolipoproteins. J Biol Chem, 2010. 285(42): p. 32486-93.

[45] Geng, X., et al., alpha-Synuclein binds the K(ATP) channel at insulin-secretory granules and inhibits insulin secretion. *Am J Physiol Endocrinol Metab,* 2011. 300(2): p. E276-86.

[46] Sudhof, T.C. and J. Rizo, Synaptic vesicle exocytosis. *Cold Spring Harb Perspect Biol,* 2011. 3(12).

[47] Braun, A.R., et al., alpha-Synuclein induces both positive mean curvature and negative Gaussian curvature in membranes. *J Am Chem Soc,* 2012. 134(5): p. 2613-20.

[48] Lee, F.J., et al., Direct binding and functional coupling of alpha-synuclein to the dopamine transporters accelerate dopamine-induced apoptosis. *FASEB J*, 2001. 15(6): p. 916-26.

[49] Sidhu, A., C. Wersinger, and P. Vernier, alpha-Synuclein regulation of the dopaminergic transporter: a possible role in the pathogenesis of Parkinson's disease. *FEBS Lett*, 2004. 565(1-3): p. 1-5.

[50] Wersinger, C. and A. Sidhu, Attenuation of dopamine transporter activity by alpha-synuclein. *Neurosci Lett,* 2003. 340(3): p. 189-92.

[51] Nemani, V.M., et al., Increased expression of alpha-synuclein reduces neurotransmitter release by inhibiting synaptic vesicle reclustering after endocytosis. *Neuron*, 2010. 65(1): p. 66-79.

[52] Perlmutter, J.D., A.R. Braun, and J.N. Sachs, Curvature dynamics of alpha-synuclein familial Parkinson disease mutants: molecular simulations of the micelle- and bilayer-bound forms. *J Biol Chem*, 2009. 284(11): p. 7177-89.

[53] Galvin, J.E., et al., Differential expression and distribution of alpha-, beta-, and gamma-synuclein in the developing human substantia nigra. *Exp Neurol,* 2001. 168(2): p. 347-55.

[54] Kumer, S.C. and K.E. Vrana, Intricate regulation of tyrosine hydroxylase activity and gene expression. *J Neurochem,* 1996. 67(2): p. 443-62.

[55] Alerte, T.N., et al., Alpha-synuclein aggregation alters tyrosine hydroxylase phosphorylation and immunoreactivity: lessons from viral transduction of knockout mice. *Neurosci Lett,* 2008. 435(1): p. 24-9.

[56] Liu, D., et al., Silencing alpha-synuclein gene expression enhances tyrosine hydroxylase activity in MN9D cells. *Neurochem Res*, 2008. 33(7): p. 1401-9.

[57] Christenson, J.G., W. Dairman, and S. Udenfriend, On the identity of DOPA decarboxylase and 5-hydroxytryptophan decarboxylase (immunological titration-aromatic L-amino acid decarboxylase-serotonin-dopamine-norepinephrine). *Proc Natl Acad Sci U S A,* 1972. 69(2): p. 343-7.

[58] Duchemin, A.-M., et al., Phosphorylation and activation of brain aromatic amino acid decarboxylase by cyclic AMP-dependent protein kinase. *J Neurochem,* 2000. 75(2): p. 725-731.

[59] Leal, R.B., et al., Tyrosine hydroxylase dephosphorylation by protein phosphatase 2A in bovine adrenal chromaffin cells. *Neurochem Res,* 2002. 27(3): p. 207-13.

[60] Janssens, V. and J. Goris, Protein phosphatase 2A: a highly regulated family of serine/threonine phosphatases implicated in cell growth and signalling. *Biochem J,* 2001. 353(Pt 3): p. 417-39.

[61] Waxman, E.A. and B.I. Giasson, Specificity and regulation of casein kinase-mediated phosphorylation of alpha-synuclein. *J Neuropathol Exp Neurol,* 2008. 67(5): p. 402-16.

[62] Lee, K.W., et al., Enhanced phosphatase activity attenuates alpha-synucleinopathy in a mouse model. *J Neurosci*, 2011. 31(19): p. 6963-71.

[63] Wu, J., et al., Lewy-like aggregation of alpha-synuclein reduces protein phosphatase 2A activity in vitro and in vivo. *Neuroscience*, 2012. 207: p. 288-97.

[64] Gong, C.X., et al., Phosphoprotein phosphatase activities in Alzheimer disease brain. *J Neurochem*, 1993. 61(3): p. 921-7.

[65] Wakabayashi, K., et al., The Lewy body in Parkinson's disease: molecules implicated in the formation and degradation of alpha-synuclein aggregates. *Neuropathology*, 2007. 27(5): p. 494-506.

[66] Berg, D., O. Riess, and A. Bornemann, Specification of 14-3-3 proteins in Lewy bodies. *Ann Neurol*, 2003. 54(1): p. 135.

[67] Nakashima, S. and F. Ikuta, Tyrosine hydroxylase protein in Lewy bodies of parkinsonian and senile brains. *J Neurol Sci,* 1984. 66(1): p. 91-6.

[68] Kawamoto, Y., et al., 14-3-3 proteins in Lewy bodies in Parkinson disease and diffuse Lewy body disease brains. *J Neuropathol Exp Neurol,* 2002. 61(3): p. 245-53.

[69] Ubl, A., et al., 14-3-3 protein is a component of Lewy bodies in Parkinson's disease-mutation analysis and association studies of 14-3-3 eta. *Brain Res Mol Brain Res,* 2002. 108(1-2): p. 33-9.

[70] Seroogy, K., et al., Cholecystokinin and tyrosine hydroxylase messenger RNAs in neurons of rat mesencephalon: peptide/monoamine coexistence studies using in situ hybridization combined with immunocytochemistry. *Exp Brain Res,* 1989. 74(1): p. 149-62.

[71] Baxter, H.C., et al., Immunolocalisation of 14-3-3 isoforms in normal and scrapie-infected murine brain. *Neuroscience*, 2002. 109(1): p. 5-14.

[72] Chu, Y. and J.H. Kordower, Age-associated increases of alpha-synuclein in monkeys and humans are associated with nigrostriatal dopamine depletion: Is this the target for Parkinson's disease? *Neurobiol Dis*, 2007. 25(1): p. 134-49.

[73] Li, W., et al., Stabilization of alpha-synuclein protein with aging and familial parkinson's disease-linked A53T mutation. *J Neurosci*, 2004. 24(33): p. 7400-9.

[74] Perez, X.A., et al., Pre-synaptic dopaminergic compensation after moderate nigrostriatal damage in non-human primates. *J Neurochem,* 2008. 105(5): p. 1861-72.

[75] Zigmond, M.J. and E.M. Stricker, Parkinson's disease: studies with an animal model. *Life Sci,* 1984. 35(1): p. 5-18.

[76] Zigmond, M.J., T.G. Hastings, and R.G. Perez, Increased dopamine turnover after partial loss of dopaminergic neurons: compensation or toxicity? *Parkinsonism Relat Disord,* 2002. 8(6): p. 389-93.

[77] Lee, V.M. and J.Q. Trojanowski, Mechanisms of Parkinson's disease linked to pathological alpha-synuclein: new targets for drug discovery. *Neuron*, 2006. 52(1): p. 33-8.

[78] Conway, K.A., et al., Accelerated oligomerization by Parkinson's disease linked alpha-synuclein mutants. *Ann N Y Acad Sci,* 2000. 920: p. 42-5.

[79] Goldberg, M.S. and P.T. Lansbury, Jr., Is there a cause-and-effect relationship between alpha-synuclein fibrillization and Parkinson's disease? *Nat Cell Biol,* 2000. 2(7): p. E115-9.

[80] Lansbury, P.T., Jr. and A. Brice, Genetics of Parkinson's disease and biochemical studies of implicated gene products. *Curr Opin Cell Biol,* 2002. 14(5): p. 653-60.

[81] Croisier, E., et al., Microglial inflammation in the parkinsonian substantia nigra: relationship to alpha-synuclein deposition. *J Neuroinflammation,* 2005. 2: p. 14.

[82] Gao, H.M., et al., Neuroinflammation and oxidation/nitration of alpha-synuclein linked to dopaminergic neurodegeneration. *J Neurosci,* 2008. 28(30): p. 7687-98.

[83] Gao, H.M., et al., Neuroinflammation and alpha-synuclein dysfunction potentiate each other, driving chronic progression of neurodegeneration in a mouse model of Parkinson's disease. *Environ Health Perspect*, 2011. 119(6): p. 807-14.

[84] Theodore, S., et al., Targeted overexpression of human alpha-synuclein triggers microglial activation and an adaptive immune response in a mouse model of Parkinson disease. *J Neuropathol Exp Neurol,* 2008. 67(12): p. 1149-58.
[85] Baba, M., et al., Aggregation of alpha-synuclein in Lewy bodies of sporadic Parkinson's disease and dementia with Lewy bodies. *Am J Pathol,* 1998. 152(4): p. 879-884.
[86] Mosharov, E.V., et al., Alpha-synuclein overexpression increases cytosolic catecholamine concentration. *J Neurosci,* 2006. 26(36): p. 9304-11.
[87] Mosharov, E.V., et al., Interplay between cytosolic dopamine, calcium, and alpha-synuclein causes selective death of substantia nigra neurons. *Neuron*, 2009. 62(2): p. 218-29.
[88] Foppoli, C., et al., Catecholamines oxidation by xanthine oxidase. *Biochim Biophys Acta,* 1997. 1334(2-3): p. 200-6.
[89] Graham, D.G., et al., Autoxidation versus covalent binding of quinones as the mechanism of toxicity of dopamine, 6-hydroxydopamine, and related compounds toward C1300 neuroblastoma cells in vitro. *Mol Pharmacol,* 1978. 14(4): p. 644-53.
[90] Hastings, T.G., Enzymatic oxidation of dopamine: the role of prostaglandin H synthase. *J Neurochem*, 1995. 64(2): p. 919-24.
[91] Linert, W., et al., Dopamine, 6-hydroxydopamine, iron, and dioxygen--their mutual interactions and possible implication in the development of Parkinson's disease. *Biochim Biophys Acta,* 1996. 1316(3): p. 160-8.
[92] Paris, I., et al., Copper neurotoxicity is dependent on dopamine-mediated copper uptake and one-electron reduction of aminochrome in a rat substantia nigra neuronal cell line. *J Neurochem*, 2001. 77(2): p. 519-29.
[93] Paris, I., et al., Dopamine-dependent iron toxicity in cells derived from rat hypothalamus. *Chem Res Toxicol,* 2005. 18(3): p. 415-9.
[94] Segura-Aguilar, J. and C. Lind, On the mechanism of the Mn3(+)-induced neurotoxicity of dopamine:prevention of quinone-derived oxygen toxicity by DT diaphorase and superoxide dismutase. *Chem Biol Interact*, 1989. 72(3): p. 309-24.
[95] Thompson, C.M., J.H. Capdevila, and H.W. Strobel, Recombinant cytochrome P450 2D18 metabolism of dopamine and arachidonic acid. *J Pharmacol Exp Ther*, 2000. 294(3): p. 1120-30.
[96] Barbeau, A., L-dopa therapy in Parkinson's disease: a critical review of nine years' experience. *Can Med Assoc J*, 1969. 101(13): p. 59-68.

[97] Fabbrini, G., et al., Levodopa-induced dyskinesias. *Mov Disord,* 2007. 22(10): p. 1379-89; quiz 1523.

[98] Del-Bel, E. and V. Tumas, *Parkinson's Disease: Some Current Therapies and their Limitations, in Frontiers in Parkinson's Disease Research,* A.S.J. Aguiar and R.D.S. Predegir, Editors. 2012, Nova Science Publishers, Inc.: New York. p. 93-116.

[99] Muangpaisan, W., et al., A systematic review of the worldwide prevalence and incidence of Parkinson's disease. *J Med Assoc Thai,* 2011. 94(6): p. 749-55.

[100] Beach, T.G., et al., Olfactory bulb alpha-synucleinopathy has high specificity and sensitivity for Lewy body disorders. *Acta Neuropathol,* 2009. 117(2): p. 169-74.

[101] Braak, H., et al., Stages in the development of Parkinson's disease-related pathology. *Cell Tissue Res,* 2004. 318(1): p. 121-34.

[102] Shannon, K.M., et al., Is alpha-synuclein in the colon a biomarker for premotor Parkinson's disease? Evidence from 3 cases. *Mov Disord*, 2012. 27(6): p. 716-9.

Index

D

E

F

M

N

O

T

U

V

W

Y